Fundamentals of

INDUSTRIAL QUALITY CONTROL

Third Edition

LAWRENCE S. AFT

CRC Press
Taylor & Francis Group
Boca Raton London New York

CRC Press is an imprint of the
Taylor & Francis Group, an informa business

Acquiring Editor: *Dennis McClellan*
Project Editor: *Christine L. Winter*
Cover Design: *Dawn Boyd*
Prepress: *Kevin Luong and Carlos Esser*

Library of Congress Cataloging-in-Publication Data

Catalog record is available from the Library of Congress.

CRC Press
6000 Broken Sound Parkway, NW
Suite 300, Boca Raton, FL 33487
270 Madison Avenue
New York, NY 10016
2 Park Square, Milton Park
Abingdon, Oxon OX14 4RN, UK

Preface

This book was written to fill the need for a basic introductory text on the fundamental concepts of statistical quality control. Many individuals who do not have a strong mathematical or statistical background need to understand the basics of the statistical methods used in quality control. The material is presented in a logical, straightforward manner.

The earlier editions of this book have been used by many individuals as part of their preparation for the ASQC certification exams. Changes from earlier editions include the further clarification of some statistical methods; additional coverage of control charts, including the s chart, the u chart, and the np chart; additional information on standard sampling plans; and amplification of some of the reliability concepts.

A word of thanks to several of my colleagues who were kind enough to review the revised manuscript and offer constructive criticism. These people include Robert Atkins, Patricia Carden, Lori Cook, and Sandra Reynolds.

The Author

Lawrence (Larry) S. Aft, P.E., has been a Professor in the Industrial Engineering Technology Department at Southern Polytechnic State University in Marietta, Georgia, since 1971. As a faculty member he has been responsible for the Quality Assurance option to the BSIET degree and the MSQA degree offered by the university. He is a Fellow of the American Society for Quality and has been actively involved with ASQ, serving on the Executive Board of the Greater Atlanta Section in many roles, including section chair, education chair, program chair, faculty advisor to the Southern Tech Student Branch of ASQ, and Chair of the Southeastern Quality Conference. He has also served as President of the Atlanta chapter of IIE and as Conference Chair for the Society for Work Science. He has presented numerous papers, written many technical articles, and authored five textbooks in quality control and industrial engineering. A registered professional engineer, Larry has provided consulting and training expertise to more than 80 organizations. He received his BSIE from Bradley University and his MS in Industrial Engineering from the University of Illinois.

Contents

1 Introduction

Introduction

The U.S., after and perhaps because of many fits and starts, has once again begun to assert itself as one of the giants of the industrial world. A major contributor to the turnaround has been the continuing emphasis on quality: quality control, quality assurance, quality improvement, and quality systems. Looking back years from now, we will probably call the 1980s and the 1990s the "decades of strategic quality management." Companies used many different names for their quality efforts: total quality management, company-wide quality improvement, continuous quality improvement, and just about any other name that included quality and the organization's name. The movement included not only industry, but service organizations, governmental organizations, healthcare, and education. These names were meant to signify a new approach to quality management that included a revolutionary rate of quality improvement in every part of the company. Organizations rallied behind making quality process improvements and providing ever-increasing value for their customers. "Quality management was used in every business process: in marketing and market research, in finance, in the patent offices, in sales and service, in shipping, in the research and development labs, and in white-collar administrative offices" (Godfrey, 1990, p. 17).

Quality has been defined in many ways. Some sources cite consistency at the target value, while others define quality as conformance to requirements or production with specification limits. Yet others define quality as meeting the customers' requirements. All of these definitions are appropriate, and all can apply in the study of basic quality control methods. If all measurable quality characteristics are on target, they will certainly be within specification

1

limits. Customer consultation when specification limits are set ensures that requirements are met.

"The quality of a manufactured product is a direct reflection of management's attitude toward quality. Just as the winning attitude of a football team or other team . . . plays an important part in the outcome of the game, the quality attitude of a company affects the integrity of the items produced" (Adamson, 1983, p. Q1). For many years the quality of products produced in the U.S. was unquestioned. "During the early 1950's, Western product quality was regarded as best" (Juran, 1982; p. 16). The Western world, as exemplified by the U.S., led the world in quality of work produced. Prior to World War II and immediately thereafter, Japanese product quality was considered to be among the poorest, if not the worst, in the world (Juran, 1982, p. 16).

This is beginning to change. To cite some specific examples: "Xerox has regained four percentage points of market share from its worldwide competitors. In one of the toughest businesses in the U.S., nuclear fuel, the Westinghouse Commercial Nuclear Fuel Division has taken over 40 percent of the U.S. market and over 20 percent of the world market. In specialty steels, a market that most companies had abandoned to foreign competitors, a small company with plants in Ohio and Alabama, Globe Metallurgical, is running 24 hours a day, seven days a week and capturing new markets throughout the world. In textiles, another threatened U.S. industry, Milliken is recording record profits. In electronics, probably the most publicized threatened industry in the U.S., Motorola is not only recording record profits, but it is world leader in cellular telephones and has even successfully entered the Japanese market with miniature pagers capturing significant market share" (Godfrey, 1990, p. 17).

According to Dr. W. E. Deming, who presented a series of seminars to the Japanese on statistical quality control, "Statistical quality control is the application of statistical principles and techniques in all stages of production directed toward the most economic manufacture of a product that is maximally useful and has a market" (Deming, 1950, p. 3). Quality control involves uniform production of a product that meets the needs of the consumer.

In the early 1950s, at about the same time that Dr. Deming was presenting his lectures on statistical quality control to Japanese engineers and managers, Dr. Joseph Juran was, literally, preaching to management the benefits of making quality control part of the overall management philosophy and strategy. As a result, the Japanese pioneered in the following areas (Juran, 1982, p. 16):

- Introduction of massive quality-related training programs
- Establishment of annual programs of quality improvement
- Development of upper-management leadership of the quality function

Obviously, some organizations have succeeded, while others have failed. The difference most frequently lies in the quality management function. Quality activities must be led by top management, which must provide the vision and direction for continuing quality management. This effort involves activities such as setting goals that are consistent with the vision, allocating the resources necessary for reaching those goals, making quality a part of the organization's strategic operation, appraising quality performance, and recognizing and rewarding quality improvement. Management is also responsible for implementing a structure that will enable the organization to achieve the goals established by the organizational vision. While no one system is appropriate for all organizations, there have emerged some model quality systems that are impacting all of American business and industry. As American business and industry have been increasingly aware of the need to establish and maintain formal quality systems, major guides for the establishment of formal quality systems have been externally mandated or suggested. These are quality systems standards such as ISO 9000, QS-9000, and the Baldrige National Quality Award standards.

Organizations wishing to establish themselves as legitimate *quality* organizations have, to varying degrees, complied with the requirements of these systems. One of the common threads which ties all of these together relates to measurement systems and the use of statistics within the operations of the organization.

ISO 9000

ISO 9000 is a generic baseline series of quality standards written with the intent of being broadly applicable to a wide range of varying industries and products. The standards define the basics of how to *establish, document,* and *maintain* an effective quality system.

ISO 9000 standards consist of both models which define specific minimum requirements for external suppliers and guidelines for development of internal quality programs.

The ISO 9000 standards were developed by the International Organization for Standardization (ISO), a Geneva-based worldwide federation of national bodies working together through technical committees. The U.S. is represented by the American National Standards Institute (ANSI). The American Society for Quality (ASQ) has specific responsibility for the ISO 9000 series of standards, also called the Q9000 Standards. ISO 9000 consists of five standards: ISO 9000, ISO 9001, ISO 9002, ISO 9003, and ISO 9004.

ISO 9000/Q9000: Quality Management and Assurance Standards — Guidelines for Selection and Use. This provides guidance on tailoring the other standards for specific contractual situations. It also provides guidance on selection of the appropriate quality assurance model (ISO 9001/Q9001, ISO 9002/Q9002, or ISO 9003/Q9003).

ISO 9000/Q9000 challenges the user to:

1. Achieve and sustain a defined level of quality
2. Provide confidence to management that the intended level of quality is being achieved and sustained
3. Provide confidence to the purchaser that the intended level of quality is being, or will be, achieved and delivered

The most recent revision of the ISO 9000 quality system standards was adopted in 1994. According to that version of the standard, the ISO/QS standards are described as follows:

ISO 9001/Q9001: Model for Quality Assurance in Design/Development, Production, Installation, and Servicing. This is for use when conformance to specified requirements is to be assured by the supplier during several stages, which may include design/development, production, installation, and servicing. It is primarily for firms that design, produce, supply, and service products.

ISO 9002/Q9002: Model for Quality Assurance in Production and Installation. This is for use when conformance to specified requirements is to be assured by the supplier during production and installation; and this is for use by firms that supply to an externally agreed-upon (contracted) specification.

ISO 9003/Q9003: Model for Quality Assurance in Final Inspection and Test. This is for use when conformance to specified requirements is to be assured by the supplier solely at final inspection and test; and this is for use by firms that buy, sort, and re-sell products.

ISO 9004/Q9004: Quality Management and Quality System Elements Guidelines. This standard describes a basic set of elements by which quality management systems can be developed and implemented. The selection of appropriate elements contained in this standard and the extent to which these elements are adopted and applied by a company depend upon factors such as market served, nature of product, production processes, and consumer needs.

1993 Structure of ISO 9000

The ISO 9000 standards also include some subparts, as listed (the list of approved standards continues to grow as time passes):

9000-1 Quality Management and Assurance: Guidelines for Selection
9000-2 Guidelines for Implementation
9000-3 Software
9000-4 Dependability Management
9001   Quality Systems Design
9002   Quality Systems Design
9003   Quality Systems Design
9004-1 Elements — Guidelines
9004-2 Service Guidelines
9004-3 Processed Materials Guidelines
9004-4 Quality Improvement
9004-5 Quality Plans
9004-6 Project Management
9004-7 Configuration Management

ISO 9000 and Traditional Quality Efforts

ISO and Total Quality

Total Quality Management is a management system that requires cultural change for most organizations. It means putting quality in front of short-term profits and cost reduction. Quality — meeting and exceeding customer requirements — must take the dominant role in every function in the company. In order for this to happen, a framework must be in place. This framework for quality systems is ISO 9000. ISO 9000 provides the foundation for a complete total quality process. It should not be viewed as an end in itself.

Quality System Requirements

The following are the general functional areas of the ISO 9001:

1. Management responsibility
2. Quality system
3. Contract review
4. Design control
5. Document control
6. Purchasing
7. Purchaser-supplied product
8. Product identification and traceability
9. Process control
10. Inspection and testing
11. Inspection, measuring, and test equipment
12. Inspection and test status
13. Control of nonconforming product
14. Corrective action
15. Handling, storage, packaging, and delivery
16. Quality records
17. Internal quality audits
18. Training
19. Servicing
20. Statistical techniques

As the outline of the ISO 9000 contents indicates, the use of **statistical techniques** is strongly encouraged by ISO 9000 for those organizations seeking registration under the standard. There are specific requirements for statistical methods. It is imporant to identify and use appropriate statistical techniques required for verification. The validity of statistical measures needs to be determined. The following statistical tools are available:

- Process capability studies
- Process control charts
- Sampling inspection
- Regression analysis
- Tests of significance
- Design of experiments
- Analysis of variance
- Scatter diagrams
- Defect reports

Statistics, as viewed by ISO 9000, is simply the handling and use of data.

QS-9000 Quality System Requirements

The goal for *Quality System Requirements QS-9000* is the development of fundamental quality systems that provide for continuous improvement, emphasizing defect prevention and the reduction of variation and waste in the automotive supply chain. QS-9000 defines the fundamental quality system expectations of Chrysler, Ford, General Motors, truck manufacturers, and other subscribing companies for internal and external suppliers of production and service parts and materials.

These companies are committed to working with suppliers to ensure customer satisfaction beginning with conformance to quality requirements and continuing with *reduction of variation and waste* to benefit the final customer, the supply base, and themselves. QS-9000 is a harmonization of Chrysler's Supplier Quality Assurance Manual, Ford's Q-101 Quality System Standard, and General Motors' NAO Targets for Excellence, with input from the truck manufacturers. ISO 9001:1994 Section 4 has been adopted as the foundation for this standard. QS-9000 goes beyond the ISO 9001 requirements.

Specific Requirements

Some of the specific requirements called out in QS-9000 which relate to measurement systems and the use of statistics within the operations of an organization are briefly described in the following. [Numbers refer to the sections in the QS-9000 (1995) standard.]

■ Section 2.1 — Suppliers shall develop specific action plans for continuous improvement in processes that are most important to the customer once those processes have demonstrated stability and acceptable capability.

■ Section 2.2 — The supplier shall identify opportunities for quality and productivity and implement appropriate improvement projects.

■ Section 2.3 — The supplier shall demonstrate knowledge of the following measures and methodologies and shall use those that are appropriate:

■ Capability indices, C_p and C_{pk}
■ Control charts
■ Cumulative sum charting (CUSUM)
■ Design of experiments (DOE)
■ Evolutionary operation of processes (EVOP)
■ Theory of constraints
■ Overall equipment effectiveness
■ Cost of quality
■ Parts per million analysis (ppm)
■ Value analysis
■ Problem solving
■ Benchmarking

The techniques listed above involve considerable use of statistical methods. Further study of the QS-9000 standard shows the following requirement:

Statistical Techniques: Procedures (Section 4.20.2)

Selection of appropriate statistical tools is required and must be included in the control plan. Basic concepts must be understood throughout the supplier's organization.

Malcolm Baldrige Award

The Malcolm Baldrige Quality Award can serve multiple purposes. The criteria can serve as the basis for evaluating organizations pursuing recognition as outstanding organizations in terms of quality systems. The award criteria can also, more importantly, serve as the framework for maintaining a *total quality/continuous improvement* effort. It is a way that many organizations have selected to institutionalize the quality improvement process. The Baldrige Award is based upon the following.

Core Values and Concepts

Customer-driven quality — Quality is judged by customers. All product and service characteristics that contribute value to customers and lead to customer satisfaction and preference must be a key focus of a company's management system.

Leadership — A company's senior leaders need to set directions and create a customer orientation, clear and visible values, and high expectations.

Continuous improvement and learning — Achieving the highest levels of performance requires a well-executed approach to continuous improvement. The approach to improvement needs to be embedded in the way a company functions. This means improvement is part of the daily work of all units, problems are eliminated at their source, and improvement is driven by opportunities to do better. Improvements are of several types:

- Enhancing value
- Reducing errors
- Improving performance
- Improving productivity

Employee participation and development — A company's success in improving performance depends increasingly on the skills and motivation of its work force. Major challenges include:

- Integration of human resource management
- Alignment of human resource management with the business plans and strategic change processes

Fast response — Faster and more flexible response to customers is now a more critical requirement. Major improvement in response time requires simplification of work organizations and work processes.

Design quality and prevention — Management should place strong emphasis on design quality, preventing problems and waste by building quality into products and services and into production and delivery processes.

Long-range view of the future — A willingness to make long-term commitments to all stakeholders.

Management by fact — The management system needs to be built upon a framework of measurement, information, data, and analysis.

Partnership development — Build internal and external partnerships to better accomplish overall goals.

Corporate responsibility and citizenship — Corporate responsibility refers to basic expectations of a company, including ethics and protection of public health, safety, and the environment. Corporate citizenship refers to leadership and support of publicly important purposes.

Results orientation — The performance system needs to be guided and balanced by the interests of all stakeholders.

Key Baldrige Terms

The Baldrige criteria use several terms that tie directly to measurement systems and the use of statistics within the operations of an organization.

Measures and Indicators refer to numerical information that quantify input, output, and performance dimensions of processes, products, and services.

Performance refers to numerical results information obtained from processes, products, and services that permits evaluation and comparison relative to goals, standards, past results, and to others. Four types of performance are addressed:

- Operational
- Product and service quality
- Customer related
- Financial

Process refers to linked activities with the purpose of producing a product or service for a customer within or outside the company.

Productivity refers to measures of efficiency of the use of resources.

Criteria

Some of the key sections of the 1997 Baldrige Award, as they relate to measurement, include the following.

Information and Analysis (Section 4)

The Information and analysis category examines the management and use of data and information to support key company processes and improve company performance. Also examined is the adequacy of a company's internal and comparative data, information, and analysis system to support improvement of the company's customer focus, products, services, and internal operations.

- Management of information and data (Section 4.1) — Describes a company's selection and management of information and data used for strategic planning, management, and evaluation of overall performance.
- Analysis and review of company performance (Section 4.3) — Describes how data related to quality, customers, and operational performance, together with relevant financial data, are analyzed to support company-level review, action, and planning.

Business Results (Section 7)

Examines a company's performance and improvement in key business areas — customer satisfaction, financial and marketplace performance, human resources, supplier and partner performance, and operational performance. Also examined are performance levels relative to competitors.

- Customer satisfaction results (Section 7.1) — Summarizes results using key measures.
- Financial and market results (Section 7.2) — Summarizes results using key measures.
- Human resource results (Section 7.3) — Summarizes human resource results, including employee development and indicators of employee well-being and satisfaction.
- Supplier and partner results (Section 7.4) — Summarizes results using key measures.

Quality Systems

The new quality systems continue to emphasize and build on what can be called traditional quality efforts. As the following indicates, many of the functions called for in the standards involve the collection and analysis of information and data.

Inspection

The quality control function is often viewed as being solely a matter of inspection. As shown in the ISO 9000 standard, this remains a major component of quality systems. There are basically three types of inspection performed: receiving, in-process, and final inspection.

Inspection upon receipt is a physical inspection of a product before a buyer authorizes payment to a vendor. Incoming inspection takes one of three forms. In 100 percent inspection, every item is inspected for specific characteristics. In sampling inspection, a predetermined random sample is examined for compliance with given specifications. If the sample is acceptable, the entire lot is acceptable. The third form can be characterized as 0 percent inspection, or accepting the material based entirely on faith.

Example 1.1

An example of acceptance sampling can be found at a paper mill that specializes in producing recycled paper and uses as raw material primarily recycled old newspapers. As these arrive via truck, they are inspected via a sampling plan as follows: The first truckload received every month is checked for excess debris (i.e., glossy paper, paper clips, bottle caps, and the like). If the amount found exceeds a specified percentage of the truckload (by weight), then the entire truckload is rejected. For the rest of the month, every truckload received from the same supplier is inspected. If the amount found is less than the specified percentage, then only every 10th truckload is checked.

Material can also be inspected at various points in the production process. Again, every unit can be inspected, a sample can be inspected, or everything can be accepted on faith.

In addition to receiving inspection and in-process inspection, the third major point for inspecting is called final inspection. This is a check of the final product to make sure that it functions as intended. As with incoming and in-process inspection, this inspection can be 100 percent, 0 percent, or based on random samples.

Defect Analysis

Defect analysis is the quality assurance function that determines, based on scrap analysis, the current number of significant defects. The accepted method of finding the number of significant defects is based on an analysis of warranty, scrap, repair, and rework costs. Organizations should maintain records of rework costs. "This method uses Pareto's Law to pinpoint the most significant costs entailed in quality maintenance in order to allow managers to correct the most sensitive problem areas" (Barocci et al., 1983, p. 5).

Data Analysis

The development of sampling plans and the analysis of rework and scrap reports demand that certain information be available. Some of the pertinent sources of information are customer orders, technical specifications, quality characteristics, and past history.

The quantity of customer orders, or a lack of customer orders, indicates the level of an organization's commitment to quality. An organization that continually receives orders and repeat orders is probably producing a product that meets the customers' needs. A lack of orders speaks to significant problems for the organization.

Technical specifications are the requirements that must be met for a product to function as intended. Every product has certain key quality characteristics. Quality control has to be given careful attention to ensure that these key characteristics are maintained so the product provides consistent and predictable performance.

Three popular clichés come to mind at this juncture:

- Those who do not learn from the past are doomed to repeat it.
- Insanity is continuing to do the same thing in the same way and expecting different results.
- If you always do what you have always done, you will always get what you always got.

Learning from mistakes and from successes is an important quality control function.

Supplier Management

A key part of any quality assurance program is maintaining close relations with the organization's suppliers. "Suppliers are responsible for the quality of their items — hardware, software, or service. This obligation is always morally binding, usually legally binding, and if contracts are written properly, contractually binding" (Hunt, 1982, p. 54). Quality assurance and quality control include witnessing, monitoring, appraising, and documenting vendor quality systems.

Shanin and Seder suggest the following functions for a vendor relations program (in Juran, 1976, p. 24-2):

1. Publishing the vendor relations policies
2. Using multiple vendors for major procurement items
3. Identifying qualified vendors
4. Communicating this information to vendors
5. Communicating changes promptly
6. Developing methods for detecting deviations from specifications
7. Helping vendors with quality problems
8. Reviewing vendor performance

Reject Analysis

Scrap and rework analysis can provide general information on operations that have quality problems. In a complete quality control program, however, these problem areas are analyzed on a preventive appraisal and failure cost basis so that:

- Responsibilities can be recommended and assigned
- Intelligent cost decisions can be made
- Savings can be documented

Establishment of Standards for Quality

Standards for comparison should be established for every measurable performance characteristic. When a characteristic is measured, the results should be compared with some expected level of performance. This expected level of performance, or standard, can take one of several forms.

Performance standards are measures of characteristics necessary or desirable for performance, such as the ability to withstand a certain load factor under certain environmental conditions. Quality control should know all the requirements for all types of measurable characteristics and should establish what the standards will be for the products the organization produces.

The organization should also set *workmanship standards*. Workmanship refers to the complete image of quality the product gives.

Calibration

From the moment any measuring device is put into use, it begins to deteriorate in accuracy. A complete gauge control and calibration program encompasses the checking or monitoring of gauges and recalibration of all measuring devices. The testing devices have to be capable of consistently measuring what should be measured.

Quality Cost Systems

The bottom line in most organizations is the bottom line — profit. All activities must be economically justified, and quality control is no exception. The classic model for quality cost summarization includes three expenditure categories (Juran and Gryna, 1980, pp. 14–16). *Prevention costs* include expenditures for the following items:

- Quality planning
- New product review
- Training
- Process control
- Data analysis
- Quality reporting
- Quality improving

Evaluation costs are those incurred in appraising the quality of the material. These include the following:

- Inspection
- Test equipment
- Materials

The third cost category is *failure costs*. Both internal and external failures must be measured and justified. *Internal failure* costs are those that increase the cost of producing a product or providing a service. Some of these are as follows.

- Scrap
- Rework
- Retest
- Downtime
- Material disposition

As long as materials are not perfect, organizations will incur certain costs that are directly attributable to *external failure*. These include items such as:

- Complaint adjustment
- Returned material

All failure costs are measurable, some more readily than others. Reduction of any of these costs is an objective of all quality control programs.

Customer Service

"Quality should be customer driven, not technology driven, production driven, or competitor driven" (Takeuchi and Quelch, 1983, p. 140). Since customers are the final authority, the following guidelines are suggested to help make the service operation effective:

- Education
- Efficiency
- Standardization
- Involvement
- Evaluation

Conclusion

The major objective of this book is to illustrate statistical quality control concepts, focusing on the statistical applications that will help to assure quality. The first section of the book introduces some of the basic statistical methodologies used in quality control. The latter portions illustrate the application of standard statistical quality control procedures — control charts, acceptance sampling, and some basic reliability concepts.

The reader must remember at all times that these are just some of the functions that a quality control or quality assurance department has traditionally fulfilled. Statistical quality control is a tool — *a means to an end*, not an end it itself. Statistical procedures can provide "uniformity with economy" (Deming, 1950, p. 10).

As can be seen upon examination of the three quality systems documents, it is imperative that individual members of organizations have knowledge of statistical methods. The balance of this book examines the application of traditional statistical methods generally associated with the quality discipline.

References

Adamson, Craig P., "Quality Starts as an Attitude," *Quality*, August, 1983.

Barocci, T.A., Klein, T.A., Sanford, D.A., and Wever, K.A., "Quality Assurance Systems and U.S. Management in the 1980's: The Experiences of Eleven High Technology Companies," Report 2-59-83 Industrial Liaison Program, Sloan School of Management, MIT, Cambridge, Massachusetts, 1983.

Deming, W.E., *Elementary Principles of the Statistical Control of Quality*, Nippon Kagaku Gijutsu Remmei, Tokyo, Japan, 1950.

"Federal Beat," *Industrial Engineering*, August, 1983, pp. 9–10.

Godfrey, B., "Strategic Quality Management," *Quality*, March, 1990, pp. 17–27.

Gunneson, A.O., "Quality's Impact on Productivity and the Economy," *Proceedings*, Bottom Line Academia Conference, Washington, D.C., April, 1992.

Hunt, R.O., "Quality from the Source," *Quality*, April, 1992.

Juran, J. (Ed.), *Quality Control Handbook*, 3rd ed., McGraw-Hill, New York, 1976.

Juran, J., "Product Quality — A Prescription for the West," *Quality*, April, 1982, pp. 16–22.

Juran, J. and Gryna, F., *Quality Planning and Analysis*, McGraw-Hill, New York, 1980.

Takeuchi, H. and Quelch., J.A., "Quality is More than Making a Good Product," *Harvard Business Review*, July-August, 1983, pp. 139–145.

2 | Descriptive Statistics

Introduction

Two types of statistics are commonly used in business: descriptive statistics and inferential statistics. Both are critical to the application of basic statistical quality control procedures. Descriptive statistics summarize information and aid in decision making. Inferential statistics help to guide decision makers in making judgments and in evaluating information. There are several types of descriptive statistics and many ways of using inferential statistics. Inferential statistics use descriptive statistics in their analysis functions.

This chapter explains the descriptive statistics used most frequently in quality control applications — the measures of central tendency, variability, and shape. Chapters 3 and 4 deal with the descriptive statistics that help decision makers to make future predictions based on past performance — probability and probability distributions. Chapters 5 and 6 discuss the inferential statistics involved in estimation and decision making — confidence intervals and tests of hypotheses.

The Nature of Descriptive Statistics

Data are always being collected: businesses are reporting tax collections, the government is always announcing some new piece of economic information, the owner of a new car is calculating how many miles per gallon she is getting, or a sports fan is trying to make some sense out of the home town team's performance. Quality control departments also collect large amounts of data — data on inspection results, data on scrap costs, and data on vendor performance. In order to be meaningful, data must be presented and summarized so as to transmit the information they represent.

Although there are many different ways to summarize data, data are most commonly summarized graphically (via a picture) or numerically (via descriptive statistics). In either case, the information analyzed is generally sample data believed to be representative of all information of a particular type. Only rarely can data be collected on an entire population.

The entire set of observations that contain the characteristic under study is the **population**, sometimes called the universe.

A randomly selected subset representative of the population from which it is selected is a **sample**.

The purpose of this book is not to cover all the descriptive statistics, but to focus on those statistics necessary for statistical quality control. These descriptive measures will be presented strictly from an *applications* viewpoint. There are many statistics texts available that do an outstanding job of developing the theoretical aspects of these statistics.

Graphical Measures

Graphical summarization of data shows the relative amounts of data that can be categorized under each of the established data classifications. The following terms are important in discussing graphical descriptive statistics:

- **Frequency** — The absolute number of times a particular value, or group of values, occurs.
- **Class** — The particular designation that makes a certain portion of data unique. Classes may consist of individual values or groups of similar values. Classes are also referred to as cells.
- **Class (or cell) limits** — The boundaries that establish the classes. These limits are specified before the data are collected. Both lower and upper limits are stated so that there is no uncertainty regarding the cell in which observed data should be classified.
- **Midpoint** — The point halfway between the class limits. The midpoint is often the value most representative of all the possible measurements being classified within the stated interval. Computationally, the midpoint is used to represent all the measures in the class; therefore, it should be expressed to the same accuracy as the data being classified.

- **Discrete** — Some measured data can take only specified values within a given range of values or, more likely, can be measured only to a certain specified degree of accuracy. For example, although time can theoretically take on any value, it can generally be measured only to the nearest hundredth of a minute. When this is the case, the data, such as the measured time, are considered to be *discrete*.

- **Continuous** — Data that can take on any value in a given range and can be measured exactly to that value are *continuous*. Although all measurements are really discrete, when the difference between successive values is extremely small, the values are considered, for all practical intent, to be continuous.

- **Frequency distribution** — The classification or grouping of data according to some observable characteristic.

The following example illustrates the construction of a frequency distribution based on sample data.

Example 2.1

Aft-Tech manufactures a variety of products, one of which is a steel plate that has a hole drilled through it. As part of a preliminary investigation, 50 of these plates were inspected in order to determine the characteristics of the hole diameters. The diameter was measured with a linear measuring device that was calibrated to read, via a digital display, to the nearest hundredth of an inch. Because of the length and width of the steel plate, the diameter cannot exceed two inches. The minimum diameter is zero inches, for the case in which the hole was not drilled in the plate.

The analyst's task is to specify the number of classes of data desired, the class limits, and the midpoint for each class.

Solution

The first possibility considered is to have each possible reading designate a separate class. Over the entire range of values for this example, from 0 to 2 inches, there would be 200 possible classes. However, this classification system, if it were used, would not provide much information. For a data summarization to be meaningful, there must be a reasonable number of classifications or cells within which the data can be tabulated. For 50 observations, a number somewhere between 5 and 10 classes is reasonable.

Aft-Tech arbitrarily decides that eight classes would be appropriate. These will cover quarter-inch divisions over the two-inch range of possible measurements. In order to ensure that all possible readings would have a classification to "call home," the following guidelines are established:

- Limits should include the largest and smallest readings.
- Limits should be specified to one further decimal place than any possible measurement.
- Limits should be halfway between possible measurements.

The special case of a diameter of zero is eliminated; only steel plates with holes will be considered. Thus, the smallest possible diameter that can be measured and recorded is .01 inch; the largest possible is 1.99 inches. Measurements were made to the nearest hundredth of an inch, so cell limits are expressed to the nearest thousandth, halfway between possible measurements. The following class limits are established:

.005	.755	1.505
.255	1.005	1.755
.505	1.255	2.005

Each of the cells has a midpoint halfway between the cell limits. For example, halfway between 0.005 and 0.255 is 0.13. These midpoints are tabulated in column 2 of Table 2.1. So that the data can be examined, the actual measurements are tallied in column 4 to show the class to which each point belongs. The limits, along with the respective midpoints and the sum of the tally of individuals within the limits, represent a frequency distribution, since the frequency for each class has been determined. Examination shows that the class bounded by 1.005 and 1.255 has the most observations.

Table 2.1

Lower Limit	Midpoint	Upper Limit	Tally
0.005	0.13	0.255	2
0.225	0.38	0.505	5
0.505	0.63	0.755	4
0.755	0.88	1.005	10
1.005	1.13	1.255	14
1.255	1.38	1.505	8
1.505	1.63	1.755	5
1.755	1.88	2.005	2

Graphical Methods

In many cases, descriptive statistics are most effective when they readily communicate the information they represent. Thus, charts such as the frequency polygon and the frequency histogram are commonly used.

Frequency polygon — A line graph of a frequency distribution.

Frequency histogram — A bar chart of a frequency distribution.

Construction of both of these graphs is relatively straightforward once the frequency distribution has been created. The graphs have two axes. The x axis shows the classification for the data, and the y axis indicates the relative or absolute frequency for each class.

The frequency polygon is a line graph in which the frequency for each class is shown at the midpoint for the class. Figure 2.1 is the frequency polygon for the hole diameters in Example 2.1.

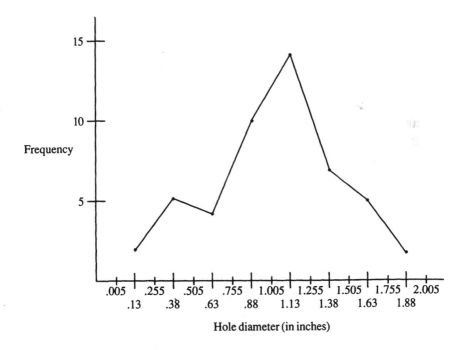

Figure 2.1 Frequency Polygon for Hole Diameter Data

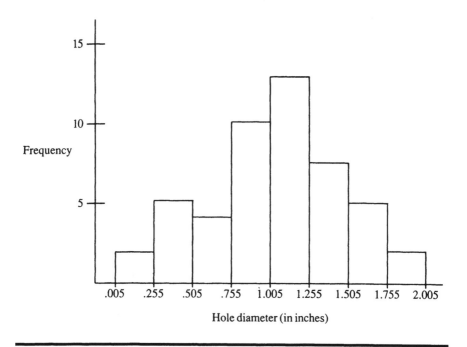

Figure 2.2   Frequency Histogram for Hole Diameter Data

The frequency histogram shows the same information as the frequency polygon but in a bolder way, as it uses a bar or rectangle that spans the entire width of the cell, from class limit to class limit. Figure 2.2 shows the frequency histogram for the hole diameters in Example 2.1.

In both charts, the largest frequency is readily observed to be in the cell with the midpoint at 1.13. Additionally, the overall shape or pattern of the data is obvious. From this shape, certain conclusions can be drawn about the distribution of hole diameters. Information can also be gained from the point at which the polygon or histogram is centered and the width of the polygon or histogram.

Histograms can be very useful in analyzing a process. At least 50 measurements should be used. Additionally, it is important to remember that the center point of the distribution may not show changes that are occurring over time. If Aft-Tech's drill were wearing and the hole size were gradually

Table 2.2

Lower Limit	Midpoint	Upper Limit	Frequency	At Least	Frequency
0.005	0.13	0.255	2	0.005	2
0.255	0.38	0.505	5	0.255	7
0.505	0.63	0.755	4	0.505	11
0.755	0.88	1.005	10	0.755	21
1.005	1.13	1.255	14	1.005	35
1.255	1.38	1.505	8	1.255	43
1.505	1.63	1.755	5	1.505	48
1.705	1.88	2.005	2	1.755	50

changing, the frequency distribution or polygon would not immediately reflect this change.

Sometimes it is useful to show the cumulative frequency in the frequency distribution and its graphical representation. It can be helpful to know how many diameters, for example, were greater than or less than a particular value. This is readily determined by accumulating frequencies. Table 2.2 gives the cumulative frequencies for the hole diameter data from Example 2.1, showing the number of plates that had hole diameters of "at least" the indicated distance. The cumulative frequency polygon and histogram for these data are shown in Figures 2.3 and 2.4, respectively. The cumulative polygon is known as an **ogive**.

Example 2.2

For the past year, the scrap control analyst at Aft-Tech has been compiling scrap costs on a weekly basis from data supplied by the metal joining department at one of the manufacturing locations. These data are listed in Table 2.3. The analyst wants to summarize the scrap data, first with a frequency distribution tabulation and then with a frequency histogram. A cumulative distribution will then be generated to show the number of weeks the scrap cost exceeded various amounts.

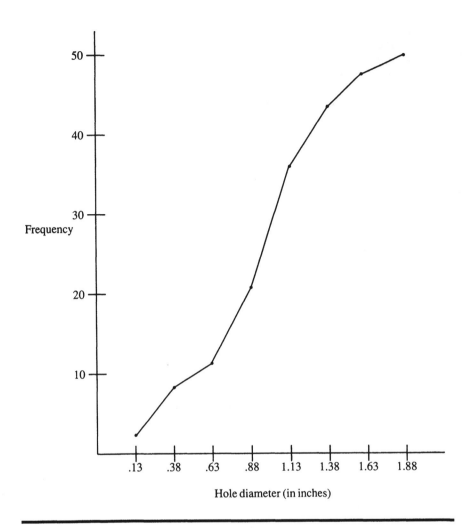

Figure 2.3   Ogive for Hole Diameter Data

Solution

An examination of the data reveals that all costs are to the nearest dollar. The analyst chooses the logical class limit of a half dollar, or 0.5, since it is halfway between actual observations. The reported scrap costs range from a minimum of $370 to a maximum of $498. The total span or range is $128, which happens to be equally divisible into 16 divisions of 8, or 8 divisions of 16.

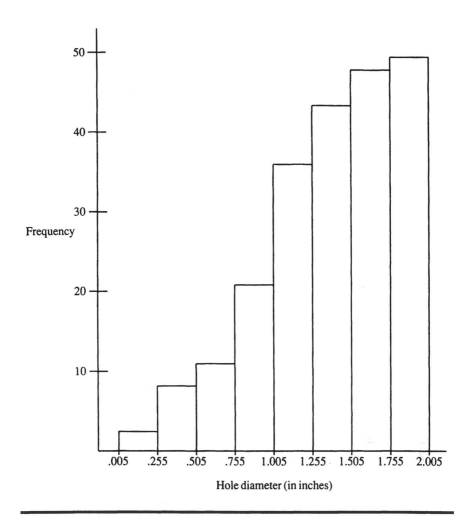

Figure 2.4 Cumulative Frequency Histogram for Hole Diameter Data

However, because money is often divided into multiples of $5, the analyst decides to use a cell width of $15. Nine classes, beginning at $370.50, will include all of the observed data.

The class limits and midpoints established for the scrap-cost data, along with the frequencies, are shown in Table 2.4. The frequency polygon and frequency histogram are shown in Figures 2.5 and 2.6, respectively.

Table 2.3

Week	Quarter 1	Quarter 2	Quarter 3	Quarter 5
1	400	416	391	409
2	450	482	419	428
3	441	421	488	447
4	473	462	427	386
5	462	453	436	465
6	412	468	476	491
7	494	447	446	380
8	495	476	464	453
9	433	434	462	434
10	427	465	455	415
11	399	456	407	496
12	452	465	498	427
13	416	445	388	486

Table 2.4

Lower Limit	Midpoint	Upper Limit	Tally
370.5	378	385.5	/
385.5	393	400.5	/////
400.5	408	415.5	////
415.5	423	430.5	///// //
430.5	438	445.5	///// /
445.5	453	460.5	///// ///
460.5	468	475.5	///// ////
475.5	483	490.5	/////
490.5	498	505.5	////

The cumulative distribution and graphs are then prepared. Cumulative frequencies, showing the number of weeks scrap exceeded the cell's lower limit, are shown in column 5 of Table 2.5. The accumulation gives the number of weeks for which the lower limit was *exceeded*. The corresponding polygon and histogram are shown in Figures 2.7 and 2.8, respectively.

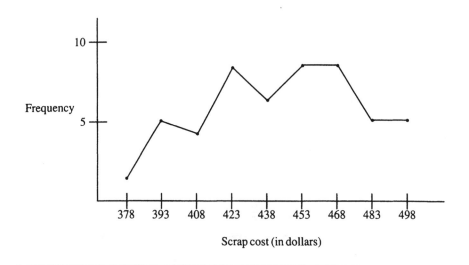

Figure 2.5   Frequency Polygon for Scrap Cost Data

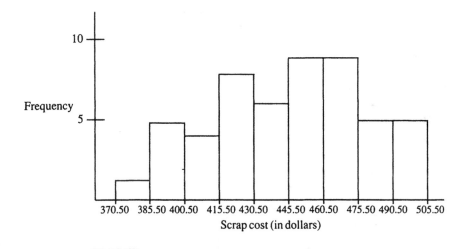

Figure 2.6   Frequency Histogram for Scrap Cost Data

Table 2.5

Lower Limit	Midpoint	Upper Limit	Frequency	Cumulative Above Lower Limit
370.5	378	385.5	1	52
385.5	393	400.5	5	51
400.5	408	415.5	4	46
415.5	423	430.5	8	42
430.5	438	445.5	6	34
445.5	453	460.5	9	28
460.5	468	475.5	9	19
475.5	483	490.5	5	10
490.5	498	505.5	5	5

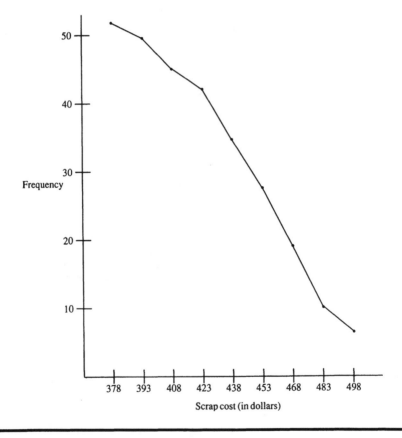

Figure 2.7   Ogive for Scrap Cost Data

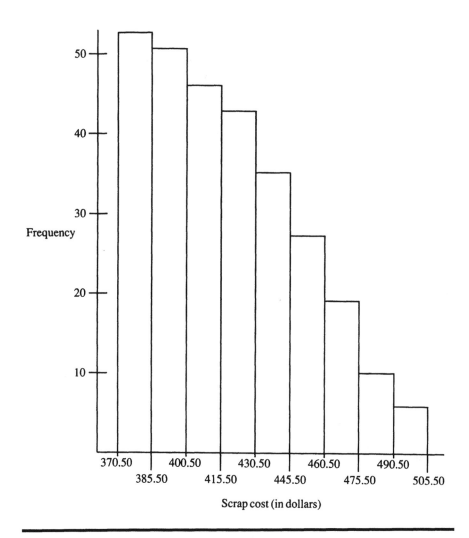

Figure 2.8 Cumulative Frequency Histogram for Scrap Data

Quantitative Descriptive Statistics

Definitions

Although pictures are useful in describing sample data, numerical statistics can sometimes transmit important information more precisely. Two general types of descriptive statistics that are important in quality control are measures of central tendency, which show the location of the data, and measures of dispersion, which indicate how widely distributed the data are.

If it were possible to obtain only one descriptive statistic for any sample, it would be most desirable to have a measure of central tendency, known generically as the average. There are several different types of averages, the most common being the arithmetic mean.

The **average** single best estimate or representative of a sample of data. The average shows the location of the sample relative to the possible values that might describe the data.

The **arithmetic mean** is the sum of the individual values in the sample, divided by the number of values in the sample.

In *fundamental* quality control applications, the arithmetic mean is the only average used. (There are, of course, a number of applications that use the other averages, such as median control charts, but these are not generally considered to be fundamental applications.) The term "average" is usually used to refer to the arithmetic mean.

Example 2.3

Inspection readings for the length of a bracket used in a warehouse-type shelf produced the following values:

$$18.02 \quad 18.01 \quad 18.07 \quad 18.05 \quad 18.04 \quad 18.03$$
$$18.05 \quad 18.00 \quad 18.05 \quad 17.97 \quad 17.99 \quad 18.02$$

Calculate the arithmetic mean for these values.

Solution

The first step is to find the sum of the individual values. This sum, designated ΣX_i, is 216.31. The sum is then divided by the total number of values in the sample, which in this case is 12.

$$\bar{x} = \frac{216.31}{12} = 18.03$$

The arithmetic mean is 18.03

The **median** is another measure of central tendency. It is the middle value of a sample, with an odd number of values or the average of the two middle values in a sample with an even number of values. It may be considered the value below which half the values in the sample fall. In order to calculate the median, the values must be in ascending or descending order.

It is most useful when a sample value is an outlier or not representative of the values of the sample.

Example 2.4

A testing lab has promised its customers that all tests will be completed within an average of 18 days. Times to completion for a sample of 8 tests are shown below:

Test 1	Test 2	Test 3	Test 4	Test 5	Test 6	Test 7	Test 8
52 days	12 days	13 days	13 days	13 days	22 days	10 days	42 days

Determine the arithmetic mean and the median for the test times.

Solution

Examination of the data indicates that the arithmetic average is

$$\frac{177}{8} = 22.13 \text{ days}$$

In order to determine the median value, the sample must be put into ascending or descending order.

$$10 \quad 12 \quad 13 \quad 13 \quad 13 \quad 22 \quad 42 \quad 52$$

The middle value in this sample will be the average of the fourth and fifth values. Because they are both 13, the average is obviously 13. The median time to complete a test at this lab is 13 days.

The difference between the arithmetic mean and the median is caused by the tests that required significantly longer to complete. An investigation of the tests indicated that each test required the sample specimens to "age" for a period of time during the testing process. With aging times sometimes in excess of 18 days, the promise of an average turnaround time should be expressed as a median instead of an arithmetic mean.

The **mode** is a third measure of central tendency. It is the value that occurs most frequently within a sample.

In Example 2.4, the mode would be 13 days.

Although the average is a valuable descriptive statistic, it is desirable to have additional information about most data samples to better describe them. The second most valuable statistic is one describing the *dispersion,* or *variability,* of a sample. In quality control, there are two important measures of variability: the range and the standard deviation.

The **range** is the difference between the largest and smallest values within a sample.

Because the range is easy to calculate, it is used to provide a quick picture of the complete spread of values within a sample. Unfortunately, it does not describe the shape of the sample distribution, but merely relates its width. A single outlier will distort the picture of the dispersion.

Example 2.5

Determine the sample range for the weights of the ten sample steel plates received from a supplier. The weights are shown in the following table.

Sample	Weight
1	10.7
2	11.4
3	14.5
4	9.5
5	15.4
6	9.1
7	16.4
8	15.4
9	13.9
10	15.1

Solution

To calculate the range, which is the difference between the largest and the smallest values in the sample, the largest individual value in the sample must be identified; in this case, 16.4. Next, the smallest individual value in the sample must be identified; in this case, 9.1. The difference between these values (16.4 − 9.1 = 7.3) is the range of this sample.

The *standard deviation* is a mathematically derived statistic that shows the average variability, or dispersion, of the individual sample values about the sample mean.

The standard deviation represents the square root of the average of the sum of the squared distances of each point from the mean. The process of taking the square root of the sum of the squares ensures inclusion of the contributions of points both larger than and smaller than the mean. It also ensures that the units of the standard deviation will be the same as the units of the original measurements. The variance, which has additional theoretical mathematical meaning, is the square of the standard deviation.

Calculations

This section provides an overview of the calculations that yield the descriptive statistics most commonly used in statistical quality control: the arithmetic mean, the range, and the standard deviation. The calculations developed here are all for sample data, as rarely will all values for a population be available as data. Some of the traditional shortcut formulas that used to be part of a complete discussion of these calculations will not be included here, as the commonplace use of electronic calculators and personal computers has made them obsolete. Virtually every spread sheet program will find the mean and standard deviation for any set of numbers.

The symbols used in the calculations are defined as follows:

Σ The total of all values within the sample is the **summation**. Sometimes a counter, represented by an i and n, are included, which, when coupled with the variable X_i, merely indicates that each value of X, ranging from the first to the nth, is counted.

X_i The variable, X_i, may assume any one of the values within the sample. A particular value within the sample is indicated by specifying the value of the subscript i. For example, the fourth value in the sample is referred to as X_4.

n   The number of values in a particular sample is the **sample size**. Sometimes this is referred to as the subgroup size.

x̄	The arithmetic mean is the **sample average.**
R	The largest value within a sample minus the smallest value within the sample is the **sample range.**
s_x	The measure of the variation of the individual values in a sample about the sample mean is the **sample standard deviation.**

The sample descriptive statistics defined above are calculated by means of the following formulas:

$$\text{Arithmetic mean} = \bar{x} = \left(\Sigma X_i\right)/n$$

$$\text{Standard deviation} = s_x = \sqrt{\left(\Sigma(X_i - \bar{x})^2\right)/(n-1)}$$

$$\text{Range} = R = X_{largest} - X_{smallest}$$

Example 2.6

Aft-Tech's welding inspector performed a destructive test on a sample of 15 welds. The following are the breaking strengths, recorded in inch pounds:

14.1	14.5	14.9	14.2	14.5	14.9	14.2	15.2
14.3	14.8	15.3	14.5	14.8	15.3	14.6	

The sample size was kept relatively small because of the cost involved in performing a destructive test. In order to describe the test results, the inspector wants to calculate the sample descriptive statistics: average, range, and standard deviation.

Solution

The sum of the 15 data points is 220.1. Since there are 15 data points, n will be 15. The average is calculated using the arithmetic-mean equation

$$\bar{x} = \frac{220.1}{15} = 14.67 \text{ inch pounds}$$

A word about precision and number of digits is appropriate at this point. It is generally good practice to report descriptive statistics to the same number of decimal places as in the original data. In this instance, Aft-Tech recorded breaking strength to the nearest 0.10 inch, therefore, the average should be reported to the same single decimal. Conventional rounding techniques should be used. In this case, the inspector should report the value of 14.7 inch pounds as the average. Although calculators and computers may show the results to eight or even more decimal places, reporting of more decimals than are found in the original data is really meaningless.

The range of the sample is the difference between the largest and smallest sample values. The largest value in this sample is 15.3, and the smallest is 14.1. Thus,

$$R = 115.3 - 14.1 = 1.2 \text{ inch pounds}$$

Remember that the range is a quick and dirty way of estimating the total dispersion, spread, or variability of a sample.

Table 2.6

X_i	$X_i - \bar{x}$	$(X_i - \bar{x})^2$
14.1	−0.6	0.36
14.2	−0.5	0.25
14.2	−0.5	0.25
14.3	−0.4	0.16
14.5	−0.2	0.04
14.5	−0.2	0.04
14.5	−0.2	0.04
14.6	−0.1	0.01
14.8	0.1	0.01
14.8	0.1	0.01
14.9	0.2	0.04
14.9	0.2	0.04
15.2	0.5	0.25
15.3	0.6	0.36
15.3	0.6	<u>0.36</u>
		2.22

Having already calculated the average, $\bar{x}$, of the sample to be 14.7 inch pounds, the inspector is now prepared to calculate the standard deviation. In order to simplify the calculations, a tabular format is used (see Table 2.6). The first column gives the value of each sample, and the second column gives the difference between each sample and the average. These differences between the values in column 1 and column 2 are squared in the third column, and the sum of the differences appears at the bottom of the column. The standard deviation formula is then used, with n = 15.

$$s_x = \sqrt{\left(\Sigma(X_i - \bar{x})^2\right)/(n-1)}$$
$$= \sqrt{(2.22/14)} = \sqrt{0.15857}$$
$$= 0.398 \text{ or } 0.4 \text{ inch pounds}$$

Types of Data

Although the calculations that have been described in this chapter can be applied to any set of data, some guidelines have been developed for the type of descriptive statistics that should be used based on the type of data collected.

Nominal data are assigned to specified groups simply on the basis of a difference in a readily observable characteristic. This is usually by name only. A part may be either classified as a "good" part or a "defective" part. "The only types of descriptive statistical methods that can be used with nominal data are the mode, frequency counts, bar charts, or any kind of one to one transformation" (Clark et al., 1974, p. 15).

Ordinal data exists when the data can be placed in categories which have a known relationship to each other. The order of that relationship can often be expressed as a greater than or less than relationship. For example, a company has 6 labor grades within its assembly department. Labor grade 6 gets paid more than grade 5 which gets paid more than labor grade 4 and so forth. Typically the median is used to describe central tendency of ordinal data. Bar graphs are often used to display ordinal data.

Interval data are similar to ordinal data, but "the scale is one for which equal units are used. If, for example, the measure goes from 1 to 50, the difference between 10 and 20 is the same as the difference between 20 and 30 because the interval of 10 is the same for both" (Clark et al.,

1974, p. 17). Typically the scale used is arbitrary, as with the measurement of temperature, using either centigrade or Fahrenheit temperature scales. Because there is a quantitative scale, arithmetic means and standard deviations are appropriate descriptive statistics for interval data. Frequency histograms and polygons can be used to display this data.

Ratio data is similar to interval data but adds a true zero point. Linear measurements are examples of ratio data. Most all descriptive statistics, including arithmetic means and standard deviations are used with ratio data.

Summary

Descriptive statistics show two major features of the sample data they describe. They show the central location of the data, and they show the spread or dispersion of the data. For individuals not well versed in statistical analysis, two graphical statistics — the frequency polygon and the frequency histogram — provide a comprehensible summary of data. The arithmetic mean and the standard deviation provide a more definitive description.

The arithmetic mean, popularly known as the average, is the most widely used quantitative descriptive statistic, with important applications in inferential statistics. However, the value of the arithmetic mean depends on every value within the sample, so one out of place value can dramatically change the average.

Reference

Clark, C. and Schkade, L., *Statistical Analysis for Administrative Decisions*, Southwestern, Cincinnati, 1974.

Practice Problems

For the inspection data in problems 1–4:

    a. Construct a frequency distribution.
    b. Plot a frequency polygon.
    c. Plot a frequency histogram.
    d. Calculate the average for the ungrouped data.

e. Calculate the range.
f. Calculate the standard deviation for the ungrouped data.

1) 18   19   24   50
    19   20   25   50
    19   21   45   61

2) .56   .72   .79   .90   .56   .73   .79   .91
    .58   .73   .80   .94   .59   .73   .81   .97
    .60   .63   .65   .68   .68   .68   .70   .71
    .71   .71   .72   .73   .74   .77   .77   .77
    .77   .77   .77   .78   .78   .78   .81   .81
    .81   .81   .82   .85   .85   .87   .88   .89
    .89   .98 1.02 1.03 1.04 1.06 1.10 1.11
   1.14 1.16 1.17 1.20

3) 12.0 12.5 13.0 13.5 13.5 13.5 14.0 14.0 14.0 14.0 14.0
    14.5 14.5 14.5 14.5 14.5 14.5 15.0 15.0 15.0 15.0 15.0
    15.0 15.0 15.5 15.5 15.5 15.5 15.5 16.0 16.0 16.0 16.0
    16.5 16.5 16.5 16.5 16.5 17.0 17.0 17.0 17.5 17.5 17.5
    17.5 17.5 18.0 18.0 18.0 18.0 18.5 18.5 18.5 19.0 19.5

4) 8   9.2   10   11.1
    8.1   9.2   10.1 11.2
    8.1   9.2   10.1 11.3
    8.2   9.2   10.2 11.4
    8.2   9.3   10.2 11.5
    8.2   9.3   10.2 11.5
    8.2   9.3   10.2 11.5
    8.3   9.3   10.2 11.6
    8.4   9.4   10.2 11.6
    8.5   9.4   10.2 11.6
    8.5   9.4   10.3 11.6
    8.5   9.4   10.3 11.7
    8.5   9.4   10.4 11.7
    8.5   9.4   10.5 11.7
    8.5   9.5   10.6 11.7
    8.6   9.6   10.6 11.7
    8.6   9.6   10.6 11.7

8.6	9.6	10.6	11.7
8.6	9.6	10.6	11.8
8.6	9.7	10.6	11.8
8.6	9.7	10.6	11.8
8.7	9.8	10.6	11.8
8.7	9.9	10.7	11.8
8.7	9.9	10.7	11.8
8.7	9.9	10.7	11.8
8.7	10	10.7	11.8
8.9	10	10.8	11.9
8.9	10	10.9	11.9
8.9	10	10.9	11.9
8.9	10	10.9	12
9	10	11	12
9	10	11	12
9	10	11	12
9.2	10	11	16.4
9.2	10	11.1	

3 Probability

Introduction

Statistical quality control is based on the premise that there is always a chance that a manufacturing process will not produce as intended. Nothing is certain to happen — the unexpected may occur at any time. The rules and theory of statistical probability permit quality control personnel to evaluate the chance that defective material will appear in a process or a product.

The relative frequency of occurrence of an event is known as it **probability.** It is the ratio of the number of favorable events to the total number of possible events.

Example 3.1

An inspector for Aft-Tech checks a sample of 1000 parts for conformance to a particular standard. The inspector determines that 40 of these do not meet the standard. What is the observed probability of a defect?

Solution

The probability that one part, selected at random, will be found to be defective is the number of favorable events, (in this case a "favorable event" is a defect), divided by the total number of possible events, or the total sample size. For 40 total defects out of a sample of 1000, the observed, or *empirical,* probability of a defect, or P(d), is

$$P(d) = 40/1000 = 0.04$$

A number of rules, or theorems, have been developed for the application of probability theory. The following section presents the definitions and rules governing probability as it applies to statistical quality control.

Probability Theorems

Several terms must be defined prior to the presentation of the theorems.

Two events are **mutually exclusive** if they have no outcomes in common or if when one occurs it keeps the other one from happening.

Within a sample of humans, being male and being female are mutually exclusive events. Similarly, evaluating a part as acceptable and evaluating it as defective are mutually exclusive events.

Two events are said to be **non-mutually** exclusive when they share a common area of occurrence or can both happen at the same time.

Within a population sample, being male and having blue eyes are not mutually exclusive events. There are some people who are males and have blue eyes.

Elementary algebraic-set theory illustrates these probability concepts best. Figure 3.1 shows the *mutually exclusive* events A and B via a Venn diagram. Note that there is no overlap of the circles representing the two events. The non-mutually exclusive events A and C are shown in Figure 3.2. Notice the area of overlap. It represents the commonality of the events.

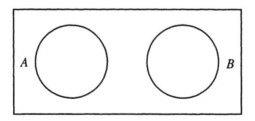

Figure 3.1   Venn Diagram for Two Mutually Exclusive Events A and B

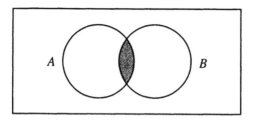

Figure 3.2   Venn Diagram for Two Non-Mutually Exclusive Events A and B

Two or more events are said to be **independent** if the occurrence of one event does *not* affect the occurrence of the second event in any way, shape, or form. Selection of one does not alter the chance of selection of the other.

The probability of selecting a defective part from a large sample of bolts is not influenced by the probability of having selected a defective part from a preceding sample of bolts.

Rules

There are two basic rules for probability. These are the *addition* and the *multiplication* rules. The addition rule is used when the probability of one event *or* another event will occur. The multiplication rule is used when the probability of one event *and* another will occur.

Addition Rule for Mutually Exclusive Events

The probability of occurrence of event A or event B, when events A and B are mutually exclusive, is the sum of the probabilities of the occurrence of event A and the occurrence of event B. Symbolically this is expressed as

$$P(A \text{ or } B) = P(A) + P(B)$$

In the language of algebra, the solution is the union of two mutually exclusive sets:

$$P(A \cup B) = P(A) + P(B)$$

Example 3.2

A sample of 100 people is selected at random from the students at State Tech. Within this sample are 61 males and 39 females. Find the probability of (a) finding a male, (b) finding a female, and (c) finding a male or a female.

Solution

(a) The probability of finding a male within this sample is 61/100 = .61.
(b) The probability of finding a female within this sample is 39/100 = .39.
(c) The probability of finding a male or a female within this sample is the sum of the probabilities:

P(male) or P(female) = P(male) + P(female) = .61 + .39 = 1.00

A probability of 1.00 has two important implications. First, it means that all of the possibilities have been suggested — the samples are *collectively exhaustive*. Second, it means that the outcome is a certainty, or a sure thing, for 1.00 is the maximum possible probability. A probability of .00 is the smallest possible probability; it means that there is no chance an event will happen. In Example 3.2 it makes intuitive sense that any individual found will be either male or female.

Example 3.3

A large bowl of beads is known to contain 400 red beads, 100 blue beads, and 500 green beads. Find the probability of selecting, in a single random sample, (a) a red bead, (b) a blue bead, (c) a green bead, or (d) a red bead or a green bead.

Solution

It is first determined that there are 1000 beads in the bowl.
(a) The probability of selecting a red bead in a single random draw is 400/1000 = .4.
(b) The probability of selecting a blue bead is 100/1000 = .1.
(c) The probability of selecting a green bead is 500/1000 = .5.
(d) If one bead is selected at random, what is the probability that it is a red *or* a green bead? Because these events are mutually exclusive the addition rule for mutually exclusive events can be applied:

P(red or green) = P(red) + P(green) = .4 + .5 = .9.

Addition Rule for Non-Mutually Exclusive Events

If two events are not mutually exclusive it is possible for both events to occur at the same time. The probability of occurrence for two non-mutually exclusive events is the sum of their probabilities minus the product of their probabilities. It is the probability of event A occurring *or* event B occurring minus the probability of event A *and* event B occurring. When both events occur it is called a *joint event.*

Symbolically this is expressed,

$$P(A \text{ or } B) = P(A) + P(B) - P(A B)$$

Example 3.4

Table 3.1 categorizes the students from State Tech as to age and gender. Determine the probability that a student selected at random from this sample is (a) male, (b) under 20, and (c) male or under 20.

Table 3.1

	Males	Females	Total
Under 20	3500	2000	5500
20 or Older	2500	2000	4500
Total	6000	4000	10000

Solution

(a) The probability that a student selected at random is male is 6,000/10,000 = .6.

(b) The probability that a student selected at random is under 20 is 5,500/10,000 = .55.

(c) The probability that a student selected at random is male or under 20 requires the use of the addition rule for non-mutually exclusive events, because these two events, being male and being under 20, are not mutually exclusive. (There can be males who are under 20 years of age. This is the joint event.)

$$P(\text{male or under 20}) = P(\text{male}) + P(\text{under 20}) -$$
$$[P(\text{male}) \text{ and } P(\text{under 20})] = (.6) + (.55) - (.35) = .80$$

Example 3.5

Determine the probability of drawing a four or a heart on a single draw from a standard deck of 52 playing cards.

Solution

These two events, drawing a four and drawing a heart, are not mutually exclusive. A four of hearts is certainly a possibility.

The probability of selecting a four is 4/52, and the probability of selecting a heart is 13/52. The chance of selecting the four of hearts is the intersection of these two events, as represented by their product:

P(four and a heart) = P(four)P(heart) = (4/52)(13/52) = 1/52

The probability of selecting a four or a heart is

P(four or a heart) = 4/52 + 13/52 – 1/52 = 4/13

Multiplication Rule for Independent Events

To determine the probability that one event will occur *and* that a second independent event will also occur requires the use of the multiplication rule. The term *and* indicates a joint probability, for which multiplication of probabilities is required.

The probability of occurrence of both an independent event A and an independent event B is the product of their respective probabilities. Remember that occurrence or nonoccurrence of one independent event does not influence the probabilities associated with other independent events.

P(A and B) = P(A)P(B)

In algebraic set theory this is the intersection of two (or more) sets:

P(A and B) = P(A B)

The probability of three independent events, A, B, and C, occurring is the product of their respective probabilities.

P(A and B and C) = P(A)P(B)P(C)

When more than one object is sampled, the effect of the sampling process on the individual events must be considered. The two basic types of sampling are replacement and nonreplacement.

In **replacement** sampling, a sample, or group of samples, is returned to the population after sampling so that the probabilities associated with selecting additional items are not changed.

The impact of non-replacement sampling is found to be more significant on smaller populations. When the population is very large relative to the sample, then the non-replacement of that sample will not alter the probabilities associated with future sampling. Replacement sampling is assumed when using the multiplication rule for independent events.

Example 3.6

What is the probability of selecting two clubs in two draws from a deck of cards with replacement?

Solution

If a card is selected at random from a playing deck there is a 13/52 chance that a club will be selected on the first draw. With replacement the same probability is present for the second selection. Thus

$$P(\text{club and club}) = (13/52)(13/52) = 169/2704 = .0625$$

Example 3.7

Consider again a deck of cards. A deck is a small sample, and if selected cards are not replaced after each draw the probabilities will change. What is the probability that hearts will be drawn on two successive draws, if the selected card is not replaced?

Solution

The probability of getting a heart on the first draw is 13/52. If a heart is selected on the first draw only 12 remain in the deck, which now contains 51 cards. The probability of getting a heart on the second draw is 12/51. Thus the probability of the joint event, getting a heart and a heart, is the product of the two events. This is $(13/52)(12/51) = .0588$.

Probability of Non-Independent Events

When two events are not independent, their probabilities are considered conditional. The occurrence of one event depends on, or is conditional upon, the occurrence of the other event. The probability of occurrence of two dependent events A and B is expressed in terms of the probability that A will occur given that B has already happened. In other words, the probability of event A is dependent on identification of event B. Symbolically, conditional probability is shown with a short vertical line:

$$P(A|B) \text{ is read, "probability of A given B"}$$

In **non-replacement** sampling, a sample, or group of samples, is not returned to the population. This changes the probabilities associated with selecting additional samples from that population.

Example 3.8

(a) What is the probability that a heart and a spade will be drawn on two successive draws, if the selected card is not replaced? (b) Determine the probability of selecting a heart and then a spade on two successive draws, if the selected card is replaced. (c) What is the probability of selecting a heart or a spade on a single draw?

Solution

(a) The probability of drawing a heart on the first draw is 13/52. Without replacement the probability of drawing a spade on the second draw is 13/51. The probability of drawing a spade, given that a heart is drawn on the first draw is the product of the two probabilities. This is $(13/52)(13/51) = .0637$.
(b) If the selected card is replaced, then the probability becomes $(13/52)(13/52) = .0625$.
(c) The chance of selecting a heart *or* a spade on one draw is the probability of selecting one independent event. The addition rule is used. The probability of either a heart or a spade is 13/52. Drawing one or the other requires the sum of the two, or $26/52 = .5$.

Example 3.9

In a machine shop, where machines 1, 2, 3, and 4 had defect rates of .02, .05, .06, and .08, respectively, find the probability that a part is defective, given that it came from machine 1.

Solution

Given that the part came from machine 1 means that the part come from a distribution that has traditionally produced material that is 2 percent defective.

Example 3.10

Given that a card selected from a deck of playing cards is a heart, what is the probability that it is a picture card?

Solution

Since there are 13 hearts and 3 picture cards the probability is 3/13.

Counting

Sometimes considerable effort is required to enumerate all possible outcomes of an event. There are just 36 possible outcomes when two dice are rolled, but the different ways defects can occur in a sample can be very large. Consider, for example, the number of different five-card poker hands possible in a 52-card deck. A number of methods that can be used to simplify the counting process are described below. Care must be taken to use the methods indicated only when the indicated conditions or desired outcomes are exactly as described.

The Multiplication Method

The multiplication or multiple choice method is used when duplication is permitted and the order in which the items appear in the sample is important. For example, consider the question of how many different five-letter combinations can be made from our alphabet. AAAAA is one, while AAAAB is a second, and AAABA is a third. To list all the possibilities would require reams of paper. The total number of different words can be determined from the formula.

$$M\binom{n}{x} = n^x$$

where n is the total sample size and X is the number in the examined subsample. In this case n = 26 letters in the alphabet and x = 5 letters in each word.

$$M\binom{26}{5} = 26^5 = 11,883,376$$

Example 3.11

Assume the following process is carried out four times: a single card is withdrawn from a deck of cards, the value and the suit are written down, the card is replaced, and the deck is shuffled. How many different arrangements of four cards is possible?

Solution

Duplication is permitted because of the replacement of the card after each draw, and the order is important in differentiating samples. The multiplication method is used, with n = 52 and x = 4.

$$M\binom{52}{4} = 52^4 = 7,311,616$$

Permutations

Permutations describe the number of possibilities when duplication is not permitted. Once an element has appeared, it may not be used again in the sample. In permutations, the order is still as important as it was in multiplication. In terms of permutations, the alphabet problem becomes one of determining how many different five letter combinations can be formed if no letter is used more than once. This means that AAAAA is no longer an acceptable arrangement. ABCDE is acceptable, though, as is ACBDE. The math formula is acceptable, as is the ACBDE. The math formula for permutations is

$$P\binom{n}{x} = n!/(n-x)!$$

For the letter example, where n = 26 and x = 5, we have

$$P\binom{26}{5} = 26!/(26-5)! = 7,893,600$$

Combinations

Combinations describe the number of possibilities when items may not be duplicated and the order is unimportant. In terms of combinations, the alphabet problem becomes one of determining how many different five letter combinations can be asked from the alphabet if any letter may only be used once and if "word" is considered unique only if it has an assortment of letters different from any other word. Thus ABCDE is one possibility and ABCDF is another, but ABCFD is the same as ABCDF. Combinations are calculated mathematically with the following relationship:

$$C\binom{n}{x} = (n!)/((x!)(n-x)!)$$

For the alphabet example, where n = 26 and x = 5 we have

$$C\binom{26}{5} = (26!)/((5!)(26-5)!) = 26!/(5!21!) = 65,780$$

Combinations are used extensively in the calculation of hypergeometric and binomial probabilities.

Example 3.12

How many different ways can a statistics class of 35 be divided into groups of 3. If the order in which the members of each group are selected is important, then the solution is as follows.

Solution

This problem fits the conditions for a permutation and the items are defined to be n = 35 and x = 3.

$$P\binom{35}{3} = (35!)/(35-3)! = 39,270$$

Example 3.13

How many different ways can an inspector select a sample of eight items from material that arrives in lots of 40?

Solution

This problem fits the case for combinations. Order is unimportant, and once an item has been selected as a sample it cannot be selected again. Since n = 40 and x = 8,

$$C\binom{40}{8} = (40!)/(8!)(40-8)! = (40!)/8!32! = 76,904,685$$

Summary

This chapter has examined the concept of probability — the chance that an ever increasing event will occur. The basic addition and multiplication rules governing empirical or observed probabilities have been explained as they relate to quality control applications. The next chapter will examine the use of probabilities for making predictions.

Practice Problems

1. A sample of 2000 items was inspected and 30 defects were discovered. What is the probability that an item selected at random from this sample will be defective?
2. Four hundred students entered State Tech last fall. Of these, 250 are still in school. What is the probability that a randomly selected student who entered school last fall is still in school this fall?
3. Aft-Tech's accounts receivable department usually manages to collect 90 percent of all outstanding bills on time. What is the probability that a random bill selected this month will be paid late?

Refer to a standard deck of playing cards for problems 4 through 8.

4. What is the probability of selecting one spade if one card is selected at random?
5. What is the probability of selecting an ace or a two or a three if one card is selected at random?
6. What is the probability of selecting an ace and a two and a three on three successive draws if the cards are replaced between selections?

7. What is the probability of selecting a seven and an eight and a nine and a ten and a jack on five successive draws if the cards are not replaced between selections?
8. What is the probability of selecting a four and a five and a six and a seven and an eight all of the same suit on five successive draws if the cards are not replaced between selections?

Table 3.2 contains information on the faculty of State Tech. Use this information for problems 9 through 18.

Table 3.2

Academic Rank	Males	Females
Assistant Professor	100	500
Associate Professor	70	270
Full Professor	430	130

9. What is the probability that a faculty member selected at random will be male?
10. What is the probability that a faculty member selected at random will be female?
11. What is the probability that a faculty member selected at random will be an assistant professor?
12. What is the probability that a faculty member selected at random will be an associate or full professor?
13. What is the probability that two faculty members selected at random will both be female?
14. What is the probability that two faculty members selected at random will both be male?
15. What is the probability that a faculty member selected at random will be either male or an associate professor?
16. What is the probability that a faculty member is a full professor, given that she is female?
17. What is the probability that a faculty member is an assistant professor, given that he is male?
18. What is the probability that a faculty member is a female, given that she is an associate professor?

Table 3.3 contains Aft-Tech's vendor inspection records for the five suppliers who normally provide a certain raw material. Use this information to answer problems 19 through 24.

Table 3.3

Supplier	Percentage Defective	Number of Lots
A	6	200
B	4	150
C	5	225
D	0.5	925
E	2	500

19. What is the probability that a lot selected at random came from vendor A?
20. What is the probability that a particular lot is from either vendor B or vendor C?
21. What is the probability that a lot is defective, given that it came from vendor D?
22. A sample of material was inspected. What is the probability that it came from vendor B?
23. The material from all the vendors is mixed in the stock area.
    (a) If 6 parts are samples, what is the probability there will be 1 defect?
    (b) If 6 parts are sampled, what is the probability that there will be 4 or fewer defects?
    (c) If 20 parts are sampled, what is the chance that there will be 17 good pieces?
24. Material from supplier E is examined. What is the probability that in a sample of 15 parts there will be 3 defects?
25. How many different outcomes are possible on three consecutive rolls of a die?
26. How many different ways can 4 board members be selected from the 20 members in an organization?
27. How many different ways can volunteers be selected from a group of 15 people to fill 6 different jobs?

4 Probability Distributions

Introduction

A **probability distribution,** or sampling distribution, is a complete listing of all possible outcomes of an experiment.

A probability distribution is a frequency distribution that shows all of the individual values within a population. Probability distributions are often expressed as mathematical relations. These equations permit the calculation of probabilities for events in the population. There are two general types of probability distributions, discrete and continuous.

A **discrete distribution** is one in which the observed characteristics fit into a finite number of categories.

For example, if products are inspected and then classified as either acceptable or unacceptable, the distribution is discrete, since all inspected products fit into one of the two categories.

A **continuous distribution** is one in which the observed characteristic may take on any value within a given range.

When electrical components are inspected for a characteristic such as electrical resistance, the distribution is continuous, since the value in ohms is limited only by the ability of the measuring equipment to record the resistance levels.

Probability distributions are extremely useful in predicting future performance based on past performance characteristics. If the products in a certain population are known, or strongly believed, to exhibit certain definable characteristics consistently, the performance of the material can be predicted from samples taken from the population. This eliminates the need to measure all of the material that any process produces.

Knowledge of the properties of various probability distributions can be quite useful. The following sections will describe several probability distributions that are important in quality control. Before these standard distributions are examined, a simple probability distribution will be developed and used to illustrate how the distributions function.

Sample Probability Distribution

In anticipation of a trip to Las Vegas, the quality control technician for Aft-Tech decides to prepare herself for the dice table. As the first step in this process, she decides to develop the probability distribution for rolling a pair of dice. The possible outcomes of rolling a pair of dice — values from 2 through 12 — represent a discrete probability distribution. Since there are multiple ways of achieving these values, the probabilities are different for each of the possible outcomes. The technician's first task in building a probability distribution is to list all of the possible outcomes and the different ways the specific events can occur.

The first event is rolling a 2. The only way this can occur is through a 1 on the first die and a 1 on the second. The second event, rolling a 3, can happen in two ways: the first die can be a 1 and the second a 2, or the first a 2 and the second a 1. Table 4.1 lists all the events, all the possibilities for each event, and the frequency of each event.

The total frequency, or total number of possibilities, is 36. Each time the pair of dice is rolled, there are 36 possible outcomes. Using the definition for probability presented in Chapter 3 — the ratio of favorable events to total possible events — the technician can determine the probability for each possible event.

The probability of rolling a 2, which is the first favorable event, is the number of favorable events, in this case 1, divided by the total number of possible events, this case 36. Thus,

$$P(2) = 1/36 = .0278$$

Table 4.1

Event	Ways to Achieve	Frequency
2	1-1	1
3	1-2; 2-1	2
4	1-3; 2-2; 3-1	3
5	1-4; 2-3; 3-2; 4-1	4
6	1-5; 2-4; 3-3; 4-2; 5-1	5
7	1-6; 2-5; 3-4; 4-3; 5-2; 6-1	6
8	2-6; 3-5; 4-4; 5-3; 6-2	5
9	3-6; 4-5; 5-4; 6-3	4
10	4-6; 5-5; 6-4	3
11	5-6; 6-5	2
12	6-6	1

All of the other possible probabilities can be similarly calculated. The various probabilities are as follows:

$$P(2) = 1/36 = .0278 \quad P(3) = 2/36 = .0556 \quad P(4) = 3/36 = .0833$$
$$P(5) = 4/36 = .1111 \quad P(6) = 5/36 = .1389 \quad P(7) = 6/36 = .1667$$
$$P(8) = 5/36 = .1389 \quad P(9) = 4/36 = .1111 \quad P(10) = 3/36 = .0833$$
$$P(11) = 2/36 = .0556 \quad P(12) = 1/36 = .0278$$

These values are normally summarized in a table, such as Table 4.2

Table 4.2

Event	Probability
2	0.0278
3	0.0556
4	0.0833
5	0.1111
6	0.1389
7	0.1667
8	0.1389
9	0.1111
10	0.0833
11	0.0556
12	0.0278
Sum	1.000

The probability distribution in Table 4.2 exhibits two key features characteristic of all probability distributions.

1. All possible outcomes are listed along with their respective probabilities.
2. The total of all the probabilities is equal to 1.0.

Now that the distribution has been developed, the technician can use it to answer the following questions about the outcomes of rolling a pair of dice. The answers, while specifically relating to the dice distribution, provide examples of the kinds of information any probability distribution can provide.

1. *What is the probability of rolling a 6 on a single roll?* This is the simplest use of a probability distribution. One needs only to look in the table for the specified event, 6, and read the corresponding probability, .1389.
2. *What is the probability of rolling a 4 or a 5 on a single play?* The addition rule applies here because *or* is used in the question. The two events are mutually exclusive, so the desired probability is the sum of the individual probabilities:

$$P(4 \text{ or } 5) = P(4) + P(5) = .0833 + .1111 = .1944$$

3. *What is the probability of rolling a 5 or lower on a single roll?* This is just a shorthand way of asking for the probability of rolling a 2 or 3 or 4 or a 5. The *or* again signifies the addition rule.

$$
\begin{aligned}
P(5 \text{ or less}) &= P\ (2 \text{ or } 3 \text{ or } 4 \text{ or } 5) \\
&= P(2) + P(3) + P(4) + P(5) \\
&= .0278 + .0556 + .0833 + .1111 \\
&= .2778
\end{aligned}
$$

4. *What is the probability of rolling a 6 or higher on a single roll?* This question can be answered two ways. As in number 2, the question can be restated to enumerate all favorable possibilities: What is the probability of rolling a 6 or a 7 or an 8 or a 9 or a 10 or an 11 or a 12? The probability is then calculated as the sum of the individual event probabilities:

$$
\begin{aligned}
P(6 \text{ or higher}) &= P(6) + P(7) + P(8) + P(9) + P(10) + P(11) + P(12) \\
&= .1389 + .1667 + .1389 + .1111 + .0556 + .0278 \\
&= .7223
\end{aligned}
$$

A second way to answer the question is to use the knowledge that the probability of rolling 6 or more is the complement of the probability of rolling 5 or less. Since the total of any probability distribution is 1.0, the desired probability can be determined by using the relationship

$$P(6 \text{ or more}) = 1 - P(5 \text{ or less}) = 1 - .2778 = .7222$$

The next sections of this chapter will examine the probability distributions most frequently found in statistical quality control applications — the hypergeometric, the binomial, the Poisson, the normal, and the exponential.

Hypergeometric Distribution

The basic probability distribution for discrete events is the hypergeometric distribution. The only condition for using the hypergeometric distribution is that the samples be random.

The **hypergeometric distribution** shows the probability of c favorable events in a sample of size n, when historically there have been d favorable events in a prior sample of size N, where $N \geq n$.

The hypergeometric distribution is useful in calculating the quality control acceptance plan probabilities associated with finding the probability of c or fewer defects in a sample of size n, when experience leads to the expectation of d defects based on a sample or population of size N.

The formula for the hypergeometric probability, where $P\left(^c_n\right)$ symbolizes the probability of c favorable events in a sample of size n, is

$$P\left(^c_n\right) = \frac{\dfrac{(N-d)!d!}{(n-c)!\big((N-d)-(n-c)\big)!(c!)(d-c)!}}{\dfrac{N!}{n!(N-n)!}}$$

Expressed in terms of combinations, the above formula is

$$P\left(^c_n\right) = \frac{\dbinom{N-d}{n-c}\dbinom{d}{c}}{\dbinom{N}{n}}$$

where

N is the prior sample size
d is the number of favorable events in the original sample
! is the symbol for factorial
n is the random sample size under consideration
c is the number of favorable events in the sample

The use of the formula for calculating hypergeometric probabilities is best illustrated through examples.

Example 4.1

What is the probability of finding 2 favorable events in a sample of 10 if a prior sample of 30 had 5 favorable events?

Solution

In this example

$$N = 30$$
$$d = 5$$
$$n = 10$$
$$c = 2$$

Thus

$$P\binom{2}{10} = \frac{\dfrac{(30-5)!(5!)}{(8!)((30-5)-(10-2))!(2!)(5-2)!}}{\dfrac{30!}{(10!)(30-10)!}}$$

$$= \frac{\dfrac{25!\ 5!}{8!(25-8)!\ 2!\ 3!}}{\dfrac{30!}{10!\ 20!}} = .36$$

There is a 36 percent chance that there will be exactly 2 favorable events in a sample of 10, if the sample is randomly selected and has the same probability distribution as the original sample.

Example 4.2

What is the probability of finding 1 favorable event in a sample of 5 if a prior sample of 20 had 2 favorable events?

Solution

Defining the terms,

$$N = 20$$
$$d = 2$$
$$n = 5$$
$$c = 1$$

Using the formula

$$
P\binom{1}{5} = \frac{\dfrac{(20-2)!(2!)}{(5-1)!\big((20-2)-(5-1)\big)!(1!)(2-1)!}}{\dfrac{20!}{(5!)(20-5)!}}
$$

$$
= \frac{\dfrac{(18!)(2!)}{4!(18-4)!(1!)(1!)}}{\dfrac{20!}{(5!)(15!)}}
$$

$$
= .3947
$$

Example 4.3

What is the probability of finding 2 or fewer favorable events in a sample of 4 if a past sample of 10 had 3 favorable events?

Solution

Defining the necessary terms.

$$N = 10$$
$$d = 3$$
$$n = 4$$
$$c = 2 \text{ or } 1 \text{ or } 0$$

For c = 2

$$P\binom{2}{4} = \frac{\dfrac{7!\ 3!}{2!\ (7-2)!\ 2!\ 1!}}{\dfrac{10!}{4!\ 6!}} = .3$$

For c = 1

$$P\binom{1}{4} = \frac{\dfrac{7!\ 3!}{3!\ (7-3)!\ 1!\ 2!}}{\dfrac{10!}{4!\ 6!}} = .5$$

For c = 0

$$P\binom{0}{4} = \frac{\dfrac{7!\ 3!}{4!\ (7-4)!\ 0!\ 3!}}{\dfrac{10!}{4!\ 6!}} = .1667$$

Using the addition rule for mutually exclusive events, we have

$$\begin{aligned}
P(2\ \text{or fewer}) &= P(2) + P(1) + P(0) \\
&= .3 + .5 + .1667 \\
&= .9667
\end{aligned}$$

The hypergeometric distribution, due to the way the probabilities are calculated, tends to be rather cumbersome to use. There also are some potential difficulties with the calculations as the sample sizes become larger.

Binomial Distribution

The **binomial distribution** is a discrete probability distribution used in defining the probability of favorable occurrences in a sample.

Because it may be used to define the probability of finding defects in a sample, the binomial distribution underlies many quality control applications. Each event must have constant probability of occurrence and each event must be independent (Juran and Gryna, 1980, p. 41). The binomial probability is defined by the expression

$$P(x) = \frac{n!}{x!(n-x)!} p^x (1-p)^{n-x}$$

where

P(x) is the probability of a favorable event, x
n is the sample size
p is the probability of a single favorable event
! is the symbol for factorial
x is the favorable event

The binomial probability distribution is most useful in describing the characteristics of populations where measurements fit an *either/or* mode. For example, a component may be acceptable or unacceptable, a tossed coin may be either a head or a tail, or some other marked distinction may differentiate between favorable and unfavorable events.

Example 4.4

Aft-Tech knows from checking its receiving inspection records that a given component supplied by a certain vendor has a 5 percent chance of being defective and a 95 percent chance of being acceptable. Upon receipt of a lot of material, a sample of 10 items is selected at random. What is the probability that this sample will contain (a) 2 defects, (b) 2 or fewer defects, and (c) more than 2 defects.

Solution

(a) Because this is an either/or situation, the binomial distribution is used to calculate the probability. The variables are identified as

$$n = 10$$
$$p = .05$$
$$x = 2$$

Thus

$$P(2) = \frac{10!}{2!(10-2)!}(.05)^2(.95)^{10-2}$$
$$= \frac{10 \cdot 9 \cdot 8 \cdot 7 \cdot 6 \cdot 5 \cdot 4 \cdot 3 \cdot 2 \cdot 1}{2 \cdot 1 \cdot 8 \cdot 7 \cdot 6 \cdot 5 \cdot 4 \cdot 3 \cdot 2 \cdot 1}(.05)^2(.95)^8$$
$$= (45)(.0025)(.6634) = .0746$$

(b) The phrase "2 or fewer" is really shorthand for 2 or 1 or 0 defects in the sample. The addition rule indicates that

$$P(2 \text{ or fewer}) = P(0) + P(1) + P(2)$$

Using the binomial distribution to find the required probabilities, we have

$$P(0) = ((10!)/(0!10!))(.05)^0(.95)^{10} = (1)(1)(.5987) = .5987$$

$$P(1) = ((10!)/(1!9!))(.05)^1(.95)^9 = (10)(.05)(.6302) = .3151$$

$$P(2) = .0746$$

So

$$P(2 \text{ or less}) = .5987 + .3151 + .0746 = .9884$$

(c) Since there must be either 2 or fewer defects or more than 2 defects in the sample, the probability complementary to that calculated in part (b) is determined.

$$P(\text{more than 2}) = 1 - P(2 \text{ or less}) = 1 - .9884 = .0116$$

Calculations involving the binomial equation can become tedious. Thus, statisticians have tabulated such values in probability tables. Table A in the Appendix is a probability table for the binomial distribution.

In order to use this table, you need three pieces of information: n, x, and p. As the following examples will illustrate, these must be specified and used in that order.

Example 4.5

Because Aft-Tech keeps thorough inspection records, it is possible to find the probability of receiving a defective product from just about any vendor. A certain nut was known to be 10 percent defective when purchased from a certain nut house. Aft-Tech's procedures call for a random sample of 20 to be inspected. Use Table A to determine the probability of finding, in the sample of 20, (a) 4 defects and (b) 4 or fewer defects.

Solution

(a) First the key information must be identified.

$$p = .10$$
$$n = 20$$
$$x = 4$$

Turning now to the binomial distribution table, we first locate the $n = 20$ section. Then the intersection of the $x = 4$ row and the $p = .1$ column specified the probability, which is .0898.

(b) Without the use of Table A, determining the probability of finding 4 or 3 or 2 or 1 or 0 defects in the sample would require a considerable amount of calculation. However, the probability can be found by summing the values for $x = 4, 3, 2, 1,$ and 0 in the $p = .1$ column of the $n = 20$ section of the table.

$$P(4) = .0898$$
$$P(3) = .1901$$
$$P(2) = .2852$$
$$P(1) = .2702$$
$$P(0) = .1216$$

The total of these is .9569.

Example 4.6

A certain product is known to be 55 percent defective. Unfortunately there is only one supplier, and the material must be purchased from that source. What is the probability that in a sample of 12 there will be (a) 2 defectives, and (b) 2 or fewer defectives?

Solution

(a) Once the key information has been identified, Table A may again be used. Given the key variables

$$p = .55$$
$$x = 2$$
$$n = 12$$

The probability is read from the table as .0093.

(b) Within the sample of 12, finding 2 or fewer defects means finding 2 or 1 or 0 defects. Using the values in Table A, we find the following probabilities:

$$.0068 + .0010 + .0001 = .0079$$

Sometimes it is desirable to use characteristics of the probability distributions for other predictive purposes. The binomial distribution, for example, is the basis for the application of a control chart known as the p chart. Although the mean and the standard deviation can be calculated by the method described in Chapter 2, there is a shortcut way to estimate the mean and standard deviation of any binomially distributed sample.

Given the sample size, n, and the probability a favorable event, p, the mean and standard deviation are calculated as follows:

$$\text{Mean} = np$$
$$\text{Standard Deviation} = \sqrt{np(1-p)}$$

When the number of binomial distributed samples, n, becomes relatively large, the binomial distribution samples, n, becomes relatively large, the binomial distribution probabilities may be approximated by the normal distribution. This distribution will be described later in this chapter.

Poisson Distribution

The **Poisson distribution** is a discrete probability distribution used to determine the probability of x occurrences in a sample of n where the probability of a favorable event is constant, but relatively small.

The Poisson distribution is often used in cases in which the probability is stated as a rate — for example, the number of defects per unit. It serves as the basis for certain control charts and for the probabilities associated with acceptance sampling.

Poisson probabilities are calculated using the probability function

$$P(x) = \frac{(np)^x e^{-np}}{x!}$$

where x is the single event, n is the sample size, p is the probability of a favorable event, and e is the base of the natural logarithms, approximately 2.718 (It is found on most calculators as the inverse of ln(x).)

Example 4.7

The textile division of Aft-Tech normally records defects in terms of the number of defects per yard of material produced. In the past, the probability of a defect has been .04. What is the probability that 20 yards of material will have (a) 2 defects and (b) 3 or fewer defects?

Solution

(a) This is a suitable application for the Poisson distribution in that it is a *per* situation. Identifying the variables for the equation,

$$n = 20$$
$$p = .04$$
$$x = 2$$
$$e = 2.718$$

The probability is then calculated:

$$P(2) = \frac{((20)(.04))^2 (2.718)^{-(20)(.04)}}{2!}$$

$$= \frac{(.8)^2 (2.718)^{-.8}}{(2)(1)} = \frac{(.64)(.4493)}{2} = .1438$$

(b) Since the probability of 3 or fewer defects is the probability of 0 or 1 or 2 or 3 defects, the Poisson probability formula is used four times, and then the results are added according to the addition rule.

$$P(0) = \frac{((20)(.04))^0 (2.718)^{-(20)(.04)}}{0!} = .4494$$

$$P(1) = \frac{((20)(.04))^1 (2.718)^{-(20)(.04)}}{1!} = .3595$$

$$P(2) = \frac{((20)(.04))^2 (2.718)^{-(20)(.04)}}{2!} = .1438$$

$$P(3) = \frac{((20)(.04))^3 (2.718)^{-(20)(.04)}}{3!} = .0383$$

$$P(3 \text{ or fewer}) = .4494 + .3595 + .1438 + .0383 = .9910$$

Like the binomial distribution table, there is a table of Poisson probabilities, reproduced in Table B of the Appendix. This table shows the probability of exactly x occurrences. In order to use Table B, it is necessary to specify n, P, and x. Table I in the Appendix is a cumulative Poisson table. It shows the probability of x or fewer occurrences. This table also requires the specification of n, p, and x.

Example 4.8

A process is known to produce material that is 6 percent defective. Material is usually sampled in samples of size 40. The process follows the Poisson distribution. What is the probability that material inspected at random will have (a) 3 defects and (b) more than 3 defects?

Solution

(a) To use the Poisson table the values of n, p, and x are identified as 40, .06, and 3, respectively. The Poisson table requires the calculation of np, or (40)(.06) = 2.4. At the intersection of the np = 2.4 column and the x = 3 row is the probability, which is .2090.

(b) More than 3 defects means 4 or 5 or 6 or..., on up to a maximum, in this instance, of 40 defects in the sample. This is a case where it is easier to work with complementary probabilities. The probability of finding more than 3 defects is the complement of the probability of finding 3 or fewer defects in the sample. The probability of finding 3 or fewer defects is the probability of finding 3 or 2 or 1 or 0 defects. Directly from the np = 2.4 column in Table B the values are as follows:

$$P(0) = .0907$$
$$P(1) = .2177$$
$$P(2) = .2613$$
$$P(3) = .2090$$

The probability of 3 or fewer is the sum of these values, or .7787. The probability of 4 or more defects is the complement, or 1 − .7787 = .2213.

Like the binomial distribution, the mean and standard deviation for the Poisson distribution are defined with equations. The mean of any Poisson distribution is equal to np. The standard deviation is the square root of the mean, or $\sqrt{np}$. With a large sample, it is advantageous to use the normal probability distribution to approximate the Poisson distribution. The normal distribution, unlike the hypergeometric, binomial, and Poisson distributions, is a continuous distribution. The distributions examined thus far are discrete distributions.

Normal Probability Distribution

The **normal probability distribution** is a continuous probability distribution used when there is a concentration of observations about the mean and equal likelihood that observations will occur above and below the mean.

Business, industrial, and service data often fit the normal probability distribution. This distribution is very useful for making predictions. As such, the study and application of statistical process control uses the distribution extensively. In a normal probability distribution the mean and standard deviation are calculated using standard formulas, and the distribution is symmetrical about the mean. Variability in the width of the distribution usually arises from many small, random (chance) and unassignable causes. For a normally distributed sample, the probability that an event will occur,

P(x), can be found from the mean and the standard deviation using the formula

$$P(x) = \frac{1}{s\sqrt{2\pi}} e^{-(x-\bar{x})^2/2s^2}$$

where

x is the favorable event
π is the constant 3.14159
e is the base of the natural logarithms, approximately 2.718
s is the standard deviation of the sample
$\bar{x}$ is the arithmetic mean of the sample

The procedure for solving this equation is beyond the scope of this book; therefore, Table C has been provided in the Appendix.

Central Limit Theorem

The normal probability distribution is very important when all of the samples of the same size are taken from a population because of the central limit theorem. This theorem implies that, without exception, *for a large sample, regardless of the shape of the distribution of the individual values within a population, sample averages selected from a population will be approximately normally distributed.* This is the key to many quality control applications of statistics. The central limit theorem permits us to operate and make decisions using samples.

Even if the frequency histogram for individuals appears as shown in Figure 4.1, the sample averages will have the histogram shown in Figure 4.2 as long as a large number of sample averages are measured.

The following relationships apply to these (and all) distributions:

$$\text{Mean}_x = \text{Mean}_{\bar{x}}$$
$$s_x = s_{\bar{x}}/\sqrt{n}$$

where

$\bar{x}$ is the distribution of sample averages
x is the distribution of individuals
s_x is the sample standard deviation
n is the size of the sample or subgroup
$s_{\bar{x}}$ is the standard deviation of the distribution of the sample averages

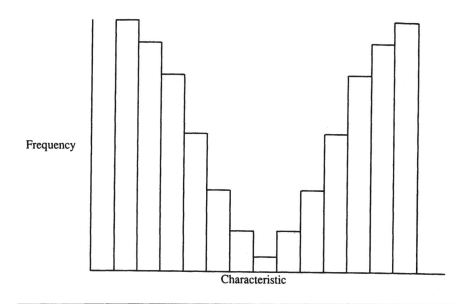

Figure 4.1 Frequency Histogram Showing Inverse Triangular Distribution

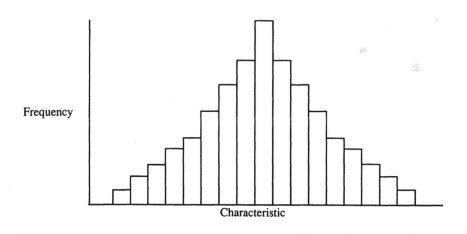

Figure 4.2 Frequency Histogram Showing Sample Averages from Samples Taken from Triangular Distribution

Standard Normal Curve

When the values for normal probabilities were first tabulated, it was necessary to build a special table for each sample based on the sample average and standard deviation. Because of the great difficulty of the mathematics involved in the procedure the values were tabulated for a special case normal distribution. This is the case of a distribution with a mean of 0 and a standard deviation of 1. This distribution is known as the *standard normal distribution*. These values can be used for any normally distributed sample by converting the sample values to the standard normal distribution via the **z transform**. Once the z transform has been calculated, the values in the standard normal table can be employed to determine normal probabilities; z is essentially the number of standard deviations the value under analysis is from the mean.

The z transform is calculated with the equation

$$z = \frac{(x - \bar{x})}{s_x / \sqrt{n}} = \frac{(x - \bar{x})}{s_{\bar{x}}}$$

where

$\bar{x}$ is the distribution of sample averages
x is the distribution of individuals
s_x is the sample standard deviation
n is the size of the sample or subgroup
$s_{\bar{x}}$ is the standard deviation of the distribution of the sample averages

After the value of z has been calculated, the corresponding probability is read from Table C. Unlike the binomial and Poisson distribution tables, the probabilities shown in this table are *not* exact probabilities. Instead, these values give the probability that an event will occur between the point, x, and the mean, x. Because the distribution is continuous, it is impossible to directly calculate probabilities of specific events. Instead, the limits of those events are defined as ranges. The standard normal probabilities are often called *areas under the normal curve*.

Example 4.9

Aft-Tech's food service operation, ever wary of the Federal Trade Commission and truth-in-packaging laws, has undertaken a study to determine the weights of the hamburgers served in the renowned company dining room.

Half pounders, it is specified, should weigh at least 8 ounces. A sample size of 500 reveals the average precooked weight to be 8.02 ounces, with a standard deviation of .4 ounces. The 8.02 is representative of the current average. The sample size of 500 approaches the population size, thus the standard deviation is assumed to be s_x.

What is the probability that a hamburger selected at random weighs (a) between 8.66 and 8.02 ounces, (b) less than 8.02 ounces, (c) less than 7.52 ounces, (d) between 7.76 and 8.26 ounces, (e) more than 9 ounces, and (f) exactly 8.61 ounces?

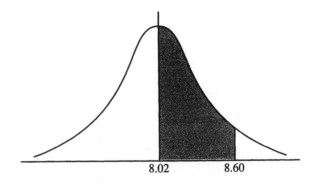

8.02   8.60

Figure 4.3

Solution

(a) It is helpful to first sketch the normal curve when attempting to answer this type of question. The shaded area in Figure 4.3 is the probability under consideration in the case of a weight between 8.66 and 8.02 ounces. The second step is to calculate the z transform so that the standard normal curve values of Table C can be used. Using the values $x = 8.66$, $\bar{x} = 8.02$, and $s_x = .4$, the z transform becomes

$$z = \frac{8.66 - 8.02}{.4} = +1.60$$

The positive sign indicates direction — that the point under consideration is larger than or to the right of the mean. A negative sign of z would indicate that the point was smaller than or to the left of the mean. For a value of 1.60 in the z column, Table C (prob column) gives a corresponding probability of .4452. Thus there is a probability of .4452 that a hamburger selected at random will have a weight between 8.66 and 8.02 ounces.

(b) A sketch is once again the first step in the solution process. The shaded area of Figure 4.4, which includes everything smaller than the mean, is the probability of interest. The figure shows that exactly half of the distribution is under consideration, and our knowledge of probability distributions tells us that half the area under the normal curve corresponds to a probability of .5. Thus there is a 50 percent chance that a hamburger selected at random will weigh less than 8.02. (Half the values will be less than the average.) Knowing this is useful in calculating other probabilities.

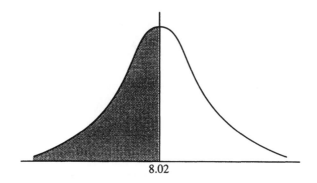

8.02

Figure 4.4

(c) Figure 4.5 includes the sketch that starts this solution process. The z value corresponding to the shaded area is calculated from $x = 7.52$, $\bar{x} = 8.02$, and $s_x = .4$.

$$z = \frac{7.52 - 8.02}{.4} = -1.25$$

The negative sign indicates that the probability is to the left of the mean. Since the desired probability is between 7.52 and negative infinity, instead of between 7.52 and the mean, an additional calculation is performed. The standard normal curve probability corresponding to the calculated z value of -1.25 is .1056, as Table C (tail column) indicates.

(d) This is a two-part problem involving the calculation of two separate probabilities: the probability of finding a weight between 7.76 and 8.02

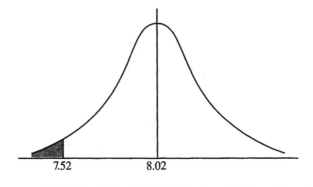

Figure 4.5

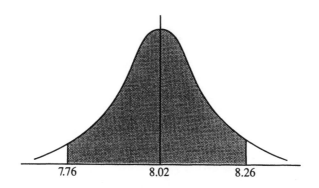

Figure 4.6

ounces and the probability of finding a weight between 8.26 and 8.02 ounces. Two z values and two normal curve values must be calculated. The initial sketch is shown in Figure 4.6. This identifies the areas corresponding to the two z values.

The first z transform is calculated from the given information:

$$z = \frac{7.76 - 8.02}{.4} = -.65$$

Table C (prob column) indicates a probability of .2422. The second z transform, corresponding to the probability of finding a value between 8.26 and 8.02, is then calculated:

$$z = \frac{8.26 - 8.02}{.4} = +.60$$

This corresponds to a probability of .2257. The total probability that a hamburger will weigh between 7.76 and 8.26 ounces is the sum of these two probabilities:

$$P(\text{between } 7.76 \text{ and } 8.26) = .2422 + .2257 = .4679$$

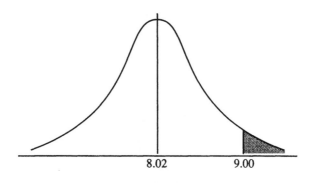

8.02   9.00

Figure 4.7

(e) The same logic used earlier to calculate the probability of finding an event between the point and negative infinity can be used here. Figure 4.7 indicates that a hamburger sampled at random will weigh more than 9 ounces. The z value is calculated based on the values of the point, mean, and standard deviation.

$$z = \frac{9.00 - 8.02}{.4} = +2.45$$

The probability from Table C (tail column) corresponding to this z value is .0071.

(f) In a continuous distribution, the specification of 8.61 ounces is an arbitrary classification. Since hamburgers can obviously be weighed to the nearest hundredth of an ounce, it would be logical to assume that any hamburger weighing between 8.605 and 8.615 ounces is called 8.61 ounces. This is really a classification system, as discussed in Chapter 2. Calculation of the probability involves two z transform calculations.

Figure 4.8 shows the area under consideration.

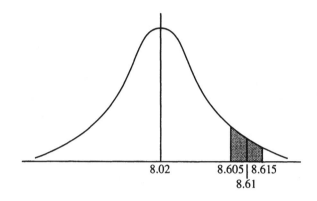

Figure 4.8

The first z value is

$$z = \frac{8.605 - 8.02}{.4} = +1.46$$

The second z value is

$$z = \frac{8.615 - 8.02}{.4} = +1.49$$

The probability that a hamburger will weigh between 8.02 and 8.605 ounces is .4279. This area is darkly shaded in Figure 4.9. The probability that a hamburger will weigh between 8.02 and 8.615 ounces is .4319. This is the entire shaded area shown in Figure 4.9. The probability of interest in this part of the example is the lightly shaded area. This probability is determined by finding the difference between the other two probabilities:

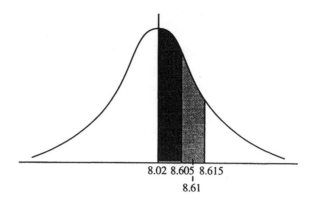

Figure 4.9

P(between 8.605 and 8.615) = .4319 − .4279 = .0040

Had the measuring devices been capable of measuring to more decimal places, a different class width would have been selected to identify a hamburger weighing 8.61 ounces. The procedure of finding the differences between the probabilities, however, would have remained the same.

Example 4.10

A cutoff operation at Aft-Tech's saw works has averaged 7.2 inches, based on samples of size 10, with a sample standard deviation of individuals of 3.79 inches. The process is believed to be normally distributed. What is the probability that a part selected at random will be (a) less than 7.4 inches, (b) 6.5 inches, and (c) between 10.8 and 3.6 inches?

Solution

(a) The probability that a part will be less than 7.4 inches corresponds to the shaded are of Figure 4.10. Because the probability is for parts less than 7.4 inches, it encompasses an area under the normal curve on both sides of the mean. First, the probability of finding a measurement between 7.2 and 7.4 inches is calculated. The z transform value is

$$z = \frac{7.4 - 7.2}{3.79/\sqrt{10}} = +.17$$

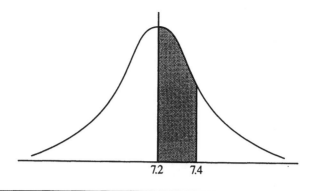

Figure 4.10

From Table C, the probability corresponding to the z value of .17 is determined to be .0675. The second part of the calculation involves finding the probability that the part will be smaller than the mean, a probability that is known to be .5. The overall probability is the sum of these two values:

$$P(\text{less than } 7.4) = .0675 + .5 = .5675$$

(b) Measurements are made to the nearest tenth of an inch, so a class *with boundaries* to one additional decimal place is developed to include all measurements that are called 6.5 inches. The cell (class) limits are 6.45 and 6.55 inches. The desired probability is shown in Figure 4.11. Once again two z transforms must be calculated. The first corresponds to the lower class limit, 6.45 and the second corresponds to the upper class limit, 6.55.

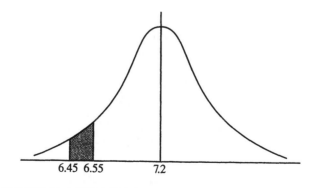

Figure 4.11

$$z = \frac{6.45 - 7.2}{1.2} = -.63$$

and

$$z = \frac{6.55 - 7.2}{1.2} = -.54$$

The probabilities for these points, as determined from Table C are .2357 and .2054. The difference between these is represented by the shaded area shown in Figure 4.11:

$$P(6.5) = .2357 - .2054 = .0303$$

(c) Figure 4.12 shows the probability that a measurement will be between 10.8 and 3.6 inches. The problem again requires the calculation of two z transform values. These are

$$z = \frac{10.8 - 7.2}{1.2} = +3.00$$

and

$$z = \frac{3.6 - 7.2}{1.2} = -3.00$$

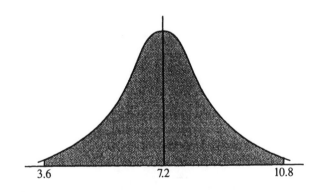

Figure 4.12

Because both z transform values are the same, the probabilities will be the same. Table C provides a value of .4987. The total probability is two times this value, or .9974. There is a 99.74 percent likelihood that a measurement will fall between values corresponding to z values of ± 3. Generally, virtually all possible values are considered to lie within z values of ± 3, which corresponds to three standard deviations from the mean.

Example 4.11

At Aft-Tech's cat food manufacturing plant, specifications call for one-pound cans to be filled with between 16.00 and 16.05 ounces of food. Analysis of sample can fill data has indicated that the weights of these cans are normally distributed, with a mean of 16.02 ounces and a standard deviation of .008 ounces (S_x). (a) What is the probability that a can selected at random will be outside the specifications? (b) If a sample of 500 cans were selected from a day's production, how many would be expected to be outside the specifications?

Solution

(a) Figure 4.13 indicates that the desired probability is for values greater than 16.05 or less than 16.00. To calculate this probability the method for complements must again be used. First the z values for areas within specifications are determined. These are as follows:

$$z = \frac{16.00 - 16.02}{.008} = -2.50$$

and

$$z = \frac{16.05 - 16.02}{.008} = +3.75$$

The respective probabilities from Table C (prob) are .4938 and .4999. The probability that a single can selected at random will be within these limits is the sum of these two probabilities:

$$P(\text{value within specs}) = .4938 + .4999 = .9937$$

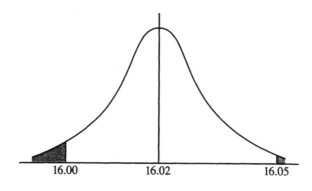

Figure 4.13

This is the complement of the area of the curve either larger than or smaller than the specifications. Thus the probability that the can will be outside the specs is

$$P(\text{value outside specs}) = 1 - .9937 = .0063$$

(b) The expected value is calculated by multiplying the sample size times the probability. Since there is a .0063 probability of not meeting the specifications, the expected value is

$$(500)(.0063) = 3.15$$

In other words, just over three cans would be either too heavy or too light.

Sometimes the normal probability distribution is used backward to determine specific values of z that correspond to desired probabilities. Example 4.12 illustrates this technique.

Example 4.12

A normally distributed process has a mean of 75 and a standard deviation of 9. (a) There is only a .05 probability of finding a value larger than what value? (b) What value of x specifies that there will be a .01 probability of randomly selecting a smaller value? (c) What symmetrical points identify a region such that there is a 95 percent chance of randomly selecting an item that falls between those points?

Solution

(a) The problem is depicted in Figure 4.14 where x is used for the unknown value. In order to determine the value of x, it is first necessary to use Table C to identify the z value that corresponds to the probability. Knowledge of the normal distribution table shows that a probability of .5 – .05 = .45 should be used. Table C shows that the z value corresponding to a probability of .45 lies between 1.64 and 1.65. Interpolation yields 1.645 for the z value. Once z is known, the original equation for the z transforms may be used, but with x as an unknown instead of z. A positive z is used because the point is larger than the mean.

$$z = \frac{x - \bar{x}}{s_x}$$
$$1.645 = (x - 75)/9$$
$$(9)(1.645) = x - 75$$
$$14.81 = x - 75$$
$$89.91 = x$$

There is a 5 percent chance, or .05 probability, of randomly selecting a value that is larger than 89.81.

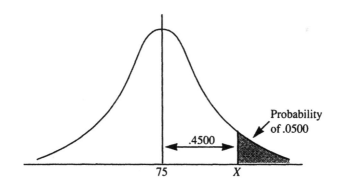

Figure 4.14

(b) Figure 4.15 illustrates the problem. Using the complementary relationship again to identify .5 – .01 = .49, as the normal distribution probability desired, Table C identified the appropriate value of z. Interpolating between 2.32 and 2.33 gives a z of 2.326 corresponding to the desired .49 probability.

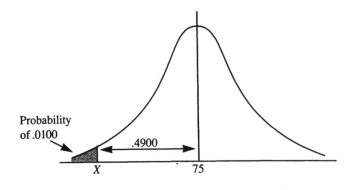

Figure 4.15

Because this value is less than the mean, a negative number for z is used to solve for x.

$$-2.326 = \frac{x - 75}{9}$$

$$(-2.326)(9) = x - 75$$

$$-20.93 = x - 75$$

$$54.07 = x$$

There is a probability of .01 that a value selected at random will be less than 54.0.

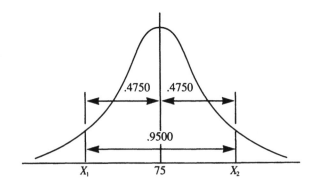

Figure 4.16

(c) Figure 4.16 depicts the situation. Because symmetrical points are specified, the total probability of .95 divides equally into two halves of .4750. Without that condition there would be an infinite number of possible points that would satisfy the problem. The z value corresponding to .4750 is 1.96. Since point x_1 is less than the mean and point x_2 is greater than the mean, both the + and the − indications are used, and the two values of x are solved for independently.

$$-1.96 = \frac{x_1 - 75}{9}$$

$$-17.64 = x_1 - 75$$

$$x_1 = 57.36$$

and

$$-1.96 = \frac{x_2 - 75}{9}$$

$$-17.64 = x_2 - 75$$

$$x_2 = 92.64$$

It would be expected that 95 percent of the values would fall between 57.36 and 92.34.

The normal probability distribution is very useful. Besides being valuable in describing data, primarily because of the properties described in the central limit theorem, it also underlies much of inferential statistics.

Exponential Distribution

The continuous **exponential distribution** is the distribution used when observations are more likely to be less than the mean than greater than the mean. (Juran and Gryna, 1980, p. 41)

The exponential distribution is most useful in describing the reliability of components. Exponential probabilities are determined by the use of the mean which is calculated from sample data. The exponential probability function follows the form

$$P(x) = e^{-x/\bar{x}}$$

where

x is the item for which the probability is desired (the favorable event)
$\bar{x}$ is the sample average
e is the base of the natural logarithms

This shows the probability that the event will be x or greater. The probability that the event will be less than or equal to x is 1 minus this probability.

The exponential distribution is continuous. The probabilities shown in probability tables, including Table D of the Appendix, are cumulative. The following examples will illustrate the calculation of exponential probabilities.

Example 4.13

Aft-Tech believes its light bulb burnout rate follows the exponential probability distribution. The mean is 600 hours and the standard deviation is 100 hours. What is the probability that a bulb selected at random will last longer than 900 hours?

Solution

Given that x = 900 and $\bar{x}$ = 600, the exponential distribution is

$$P(900 \text{ or more}) = e^{-900/600} = .2231$$

(The value of e^{-x} is read from the exponential probability distribution table, Table D.)

Example 4.14

Aft-Tech has contracted to build a radar station. The average life of a radar station without maintenance is 2000 hours. (a) What is the probability that a station will last at least 2000 hours without maintenance? (b) What is the probability that a station will last at least 250 hours without failing?

Solution

(a) The probability of lasting 2000 hours or more is calculated as follows:

$$\bar{x} = 2000 \quad x = 2000$$

$$e^{-2000/2000} = e^{-1} = .3679$$

(b) The exponential probability is calculated from the mean, 2000 hours, and a given value, 250 hours.

$$P(250 \text{ or more}) = e^{-250/2000} = .8825$$

Summary

Probability distributions provide useful information about samples and the populations from which the samples derive. Once the distribution is known, probabilities that certain events will occur can be determined and predictions can be made.

The hypergeometric distribution gives the probability of exactly a defined number of occurrences in n trials from a lot of N items having a known number of defects. This distribution can be used as long as the sample is random.

The binomial distribution assumes that the probability of occurrence, p, for each event is constant on each sample or trial. The binomial distribution is typically used when the population size is at least 10 times the sample size.

The Poisson distribution is most frequently used in determining probabilities associated with sampling methods. As with the binomial, the Poisson distribution requires that the probability of occurrence is constant on each sample. For effective use the sample size should be at least 16, the population size should be at least 10 times the sample size, and the probability of occurrence should be relatively small.

References

Juran, J. and Gryna, F., *Quality Planning and Analysis*, McGraw-Hill, New York, 1995.
Juran, J. (Ed.), *Juran's Quality Control Handbook*, New York: McGraw-Hill, New York, 1988.

Practice Problems

*In problems 1–4, use a binomially distributed sample with p = .2 and n = 14
to determine the probability of each event.*

1. x is equal to 1.
2. x is less than 3.
3. x is greater than 4.
4. x is between 2 and 5.
5. Calculate the mean and the standard deviation for a binomially dis-
    tributed sample with p = .2 and n = 14.

*In problems 6–10 use a binomially distributed sample with p = .6 and n = 10
to determine the probability of each event.*

6. x is equal to 2.
7. x is greater than 2.
8. x is less than 2.
9. x is greater than 6.
10. x is between 0 and 3.

*In problems 11–14 use a binomially distributed sample with p = .45 and n =
20 to determine the probability of each event.*

11. x is greater than 8.
12. x is less than 5.
13. x is equal to 9.
14. x is between 2 and 7.

*In problems 15–18 given that a sample follows the Poisson distribution with a
mean of 6, determine the probability of each event.*

15. x is equal to 1.
16. x is greater than 4.
17. x is less than 3.
18. x is between 4 and 6.

In problems 19–22, given that a sample follows the Poisson distribution and has a mean of .95, determine the requested value.

19. The standard deviation.
20. The probability that x is greater than 4.
21. The probability that x is less than 2.
22. The probability that x is between 1 and 4.

In problems 23–32, determine the probability of each event for a normally distributed process with a mean of 120 and a standard deviation of 6.

23. x is greater than 130.
24. x is less than 115.
25. x is greater than 118.
26. x is less than 125.
27. x is greater than 110.
28. x is between 90 and 115.
29. x is between 125 and 140.
30. x is between 128 and 132.
31. x is between 116 and 128.
32. x is equal to 119.

In problems 33–38 determine the probability of each event for a normally distributed process with a mean of .87 and a standard deviation of .072.

33. x is greater than .89.
34. x is less than .75.
35. x is equal to .82.
36. x is equal to .87.
37. x is greater than .70.
38. x is less than .65.

In problems 39–41 determine the probability of each event for an exponentially distributed process with a mean of 8.2.

39. x will be at least 2.
40. x will be at least 10.
41. x will be no greater than 6.

In problems 42–44 determine the probability of each event for an exponentially distributed process with a mean of 7800.

42. x will be at least 130.
43. x will be at least 120.
44. x will be no greater than 140.

In problems 45–49 consider a normally distributed process with a mean of 250 and a standard deviation of 25.

45. Determine the points between which 90 percent of the individual values would be found.
46. Determine the points between which 95 percent of the individual values would be found.
47. Determine the points between which 99 percent of the individual values would be found.
48. Determine the points between which 99.74 percent of the individual values would be found.
49. Determine the points between which 90 percent of averages would be found, if sample averages are calculated using samples of size four.
50. A process has a mean of .0092 and a standard deviation of .0022. If the process is normally distributed and has a lower specification of .003 and an upper specification of .0110, how many units in a sample of 800 will not meet specifications?
51. A manufacturing process has a mean of 100 and a standard deviation of 5.5. What is the probability that a value selected at random will be
    (a) greater than 106?
    (b) less than 99?
    (c) between 90 and 110?
    (d) equal to 100?
    (e) equal to 97?
52. Service times at a restaurant are believed to be normally distributed with a mean of 12.4 minutes and a standard deviation of 1.6 minutes. What is the probability that a customer will be served
    (a) in less than 10 minutes?
    (b) in more than 15 minutes?
53. Based on 100 samples of 9, breaking strength of steel springs has averaged 18.77 with a standard deviation of 1.39. What is the probability that the breaking strength of a spring selected at random will be

(a) greater than 20?
(b) less than 15.55
(c) less than 19.43?
(d) greater than 14.69?
(e) between 14.5 and 18.5?

54. Based on samples of 16, resistance of resistors produced by a certain manufacturer average 988 ohms with a standard deviation of 15 ohms. What is the probability that a resistor selected at random from this manufacturer's product will have a measured resistance of
(a) less than 1000 ohms?
(b) greater than 950 ohms?
(c) between 995 and 1010 ohms?

55. Twenty time study students studied the same operation. They discovered that 15 percent of the time was spend on personal time. What is the probability that if each of the 20 students made a single observation of this operation that the worker would be found drinking coffee (personal time) during three of the observations?

56. Material has been found to be 8 percent defective in the past. What is the probability that an inspector, upon examining 15 units, will find
(a) exactly 4 defects?
(b) more than 4 defects?
(c) fewer than 4 defects?

57. The arrival rate for material at the inspection station has been determined to be distributed according to the Poisson distribution with an average arrival rate of 14 orders per hour. What is the probability that in a given hour there will be
(a) exactly 12 arrivals?
(b) more than 10 arrivals?
(c) fewer than 15 arrivals?
(d) fewer than 20 arrivals?

58. The failure rate for an electronic component follows the exponential distribution with a mean of 5568 hours. What is the probability that a component will last
(a) at least 5000 hours?
(b) no longer than 6000 hours?

5 Confidence Intervals

Introduction

Chapters 2 through 4 dealt with descriptive statistics. Samples of material were inspected, and the measured characteristics were reported and summarized. In some instances, it was discovered that observed characteristics followed known patterns, and certain predictions could be made about the likelihood that certain events would occur. A question that should arise in every analyst's mind at this point is, "Just how good is this summary information?" Inferential statistics, which will be introduced in this and the following chapter, will address this question.

Confidence

Inferential statistical analysis is the process of sampling characteristics from larger populations, summarizing those characteristics or attributes, and drawing conclusions or making predictions from the summary or descriptive information. For example, when an organization receives a shipment of steel rods, it might inspect a relatively small sample of rods for characteristics such as tensile strength or hardness. Based on the results of this sample inspection, the organization would draw conclusions, or make inferences, about the acceptability or usefulness of the entire shipment of steel rods.

It is very common in quality control to sample from a population. And since statistical quality control programs are concerned with monitoring and maintaining product consistency, it is only natural that inferential statistics should be used.

When inferences are made based on sample data, there is always a chance that a mistake will be made. The probability that the inference will be correct is referred to as the *degree of confidence* with which the inference can be stated. There are two types of mistakes that can occur: type I errors and type II errors.

Stating that the results of sampling are unacceptable when in reality the population from which the sample was taken meets the stated requirements is a **type I error.**

The probability of a type I error — rejecting what should be accepted — is known as the α risk, or *level of significance.* A level of significance of 5 percent corresponds to a 95 percent chance of accepting what should be accepted. In such an instance, the analyst would have 95 percent confidence in the conclusions drawn or the inferences made. Another interpretation would be that there is a 95 percent chance that the statements made are correct.

Stating that the results of sampling are acceptable when in reality the population from which the sample was taken does not meet the stated requirements is a **type II error.**

The probability of a type II error — accepting what should be rejected- is known as the β risk. It is very important when acceptance sampling plans are developed and used.

Introduction to Confidence Intervals

It is possible to estimate population parameters, such as the mean or the standard deviation, based on sample values. Naturally, how good the predictions are depends on how accurately the sample values reflect the values for the entire population. If a high level of confidence in the inference is desired, a large proportion of the populations should be observed. In fact, in order to achieve 100 percent confidence, one must sample the entire population. Because of the economic considerations typically involved in inspection, the selection of an acceptable confidence interval is usually seen as a trade-off between cost and confidence. Typically, 90, 95, and 99 percent confidence levels are used, with the 99.73 percent level used in certain quality control applications.

If one desired to estimate, for example, the mean of a population, the ideal plan would be to measure every member of that population and then calculate the mean. Since this is not usually practical, a sample is generally measured and a sample mean calculated. This sample mean is called a *point estimate* of the statistic, because it is a single point, or value. How good is this estimate? It is sometimes difficult to answer this question. Although the point estimate maybe the best single estimate, no definitive statement of confidence can be made about it.

In addition to stating the point estimate, it is often desirable to establish an interval within which the true population parameter may be expected with a certain degree of confidence to fall. For example, after measuring the tensile strength of steel rods, one might say that the best estimate of the average tensile strength is 732; the true mean is between 725 and 739. (If more confidence were desired for the same data, a wider interval would be specified. A 95 percent confidence interval would be even wider — for example, between 722.49 and 741.51.)

A **confidence interval** is a range of values that has a specified likelihood of including the true value of a population parameter. It is calculated from sample calculations of the parameters.

There are many types of population parameters for which confidence intervals can be established. Those important in quality control applications include means, proportions (percentages), and standard deviations.

Confidence intervals are generally presented in the following format:

Point estimate of the population parameter ±
(Confidence factor) (Measure variability) (Adjusting factor)

A number of formulas have been developed for particular cases. The remaining sections of this chapter will present, illustrate, and interpret these formulas as they relate to quality control.

Confidence Intervals for Means

The way a confidence interval for means is developed depends on the sample size. As noted earlier, the larger the sample size, the better the estimate. Statistically, large samples are arbitrarily defined as those with 30 or more members. Small samples have fewer than 30 members.

Either a z value or a t value is needed to represent the level of significance in confidence interval formulas when one is dealing with averages. Other probability distributions are used when dealing with confidence intervals for other statistics, such as standard deviations.

z Values

For large samples, the standard normal curve is used to specify the confidence. The value corresponding to the confidence is the z value.

Example 5.1

What z value corresponds to an α of .05 or 95 percent confidence?

Solution

We are looking at a confidence interval for which the area outside the interval is symmetrically distributed. As shown in Figure 5.1, each tail of the normal curve includes $\alpha/2$, or .025, of the total area; .4750 of the total area falls between z and the mean. Table C in the Appendix indicates that the z value corresponding to .0250 is 1.96.

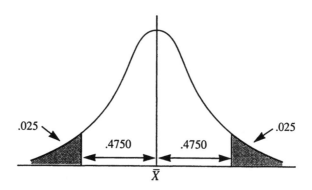

Figure 5.1

---

t Values

The student t distribution (or the t distribution) can be used with small samples. Table E in the Appendix gives the student t distribution. To use the student t distribution, one must know the level of significance, α, as well as the degrees of freedom, which is defined for each type of parameter that is estimated. For sample means, the degrees of freedom, or df, is equal to the sample size, n, minus 1. Table E lists both one-tailed α values for cases in which the "tail" outside the interval is entirely on one side of the curve, and two-tailed α values, for cases in which the area outside the interval is symmetrically distributed. For confidence intervals, a two-tailed α is usually used.

Example 5.2

What t value corresponds to an α of .05 or 95 percent confidence, and 15 degrees of freedom?

Solution

The area outside a confidence interval is symmetrically distributed on each side, so the two tailed α line of the student t table is used. For α = .05 and df = 15, t = 2.131.

Large Samples

When sample means are taken from *large samples,* the formula for the interval estimate for means is

$$\bar{x} \pm \frac{z_{\alpha/2}s_x}{\sqrt{n}}$$

where

$\bar{x}$ is the sample average
$z_{\alpha/2}$ is the value corresponding to the specified level of significance
s_x is the sample standard deviation
n is the sample size

Example 5.3

Aft-Tech's chief inspector has determined that the average weight of the gold plating used on a sample of 100 electrical contacts was 75 milligrams, with a sample standard deviation of 6 milligrams. What is the 90 percent confidence interval for the true value of the mean?

Solution

In order to use the formula presented above, the following information must be identified:

$$\bar{x} = 75 \quad n = 100 \quad z_{\alpha/2} = z_{.05} = 1.645 \text{ (from Table C)}$$

The true value of the average weight of the gold plating can be said, with 90 percent confidence, to be between the values

$$75 + \frac{(1.645)(6)}{\sqrt{100}} = 75.99$$

and

$$75 - \frac{(1.645)(6)}{\sqrt{100}} = 74.01$$

Small Samples

When sample means are taken from *small samples* the formula for the interval estimate for means is

$$\bar{x} \pm \frac{t_{\alpha/2} s_x}{\sqrt{n}}$$

where

$\bar{x}$ is the sample average
$t_{\alpha/2}$ is the value corresponding to the specified level of significance
s_x is the sample standard deviation
n is the sample size with $df = n - 1$

Example 5.4

Aft-Tech wants to determine the breaking strength of a certain component. Ten of the components are inspected in order to determine the breaking point. The average of this sample is 18 with a sample standard deviation of

1.5. Between what limits can Aft-Tech be 99 percent confident that the true breaking strength falls?

Solution

Although the single best estimate of the breaking strength is the point estimate of 18, an interval estimate affords a better indication of the true values.

Because this sample is small, the t distribution must be used for the confidence factor. A confidence level of 99 percent corresponds to a two tailed α of .01. For df = n – 1 = 10 – 1 = 9 a t value of 3.250 is read from Table E. The confidence interval is calculated as follows:

$$18 + \frac{(3.250)(1.5)}{\sqrt{10}} = 19.54$$

$$18 - \frac{(3.250)(1.5)}{\sqrt{10}} = 16.46$$

Aft-Tech can be 99 percent confident that the true value of the mean is between 16.46 and 19.54. Had Aft-Tech wanted to be only 95 percent confident of the true value of the mean, the interval would be smaller. When the level of significance is .05, corresponding to 95 percent confidence, the $t_{\alpha/2}$ value, read from Table E, is 2.262. The resulting confidence interval becomes

$$18 + \frac{(2.265)(1.5)}{\sqrt{10}} = 19.07$$

$$18 - \frac{(2.265)(1.5)}{\sqrt{10}} = 16.93$$

Confidence Intervals for Proportions

Large Samples

For large samples, the formula for the interval estimate for proportions, which follows the binomial distribution, is

$$p \pm (z_{\alpha/2})(s_p)$$

where

> p is the proportion of favorable events in the sample
> $z_{\alpha/2}$ is the value corresponding to the level of significance
> s_p is the sample standard deviation of the binomially distributed
>   sample

Example 5.5

Aft-Tech received a shipment of 20,000 light bulbs for use in its many business locations. Of these bulbs, 400 were inspected in order to estimate the proportion defective. There were 12 bulbs in the sample of 400 that did not light when tested. Determine, with 99 percent confidence, the true percentage of defective bulbs in this population of light bulbs.

Solution

The proportion defective, p, must first be calculated:

$$p = \frac{\text{favorable events}}{\text{sample size}}$$

$$p = 12/400 = .03$$

This proportion is used to calculate the sample standard deviation:

$$s_p = \sqrt{(p)(1-p)/n} = \sqrt{(.03)(.97)/400} = \sqrt{.00007275} = .00853$$

For this two-sided situation, the z value corresponding to 99 percent confidence, an α of .01, is found in the interpolation of the values in Table C, to be 2.575.

The confidence interval is calculated

$$.03 + (2.575)(.00853) = .03 + .022 = .052$$
$$.03 - (2.575)(.00853) = .03 - .022 = .008$$

Based on this sample of 400 bulbs, it can be expected, with 99 percent confidence, that the true proportion of defective bulbs is between .052 and .008.

Example 5.6

Upon examining the output of a particular process, Aft-Tech discovered that 15 percent of the last 50 lots produced were defective. Calculate the 95 percent confidence interval for the proportion of lots produced by this process that will not meet specifications.

Solution

The sample proportion has already been determined to be .15. The sample standard deviation is then calculated:

$$s_p = \sqrt{(.15)(.85)/(50)} = \sqrt{.00255} = .0505$$

The z value corresponding to 95 percent confidence, an α of .05, is 1.96 for this two-sided confidence interval. The 95 percent confidence interval is calculated as follows:

$$.15 + (1.96)(.0505) = .2490$$
$$.15 - (1.96)(.0505) = .0510$$

Small Samples

For small samples the formula for the interval estimate for proportions is

$$p \pm t_{\alpha/2}(s_p)$$

where

> p is the proportion of favorable events in the sample
> $t_{\alpha/2}$ is the value corresponding to the level of significance
> s_p is the sample standard deviation for the binomially distributed sample

The degrees of freedom is equal to the sample size, n, minus 1.

Example 5.7

A college professor, upon completion of a new course, noted that the grades of the 9 students enrolled in the course averaged 94 percent. She wanted to find out what the true average for a large number of students taking the course might be.

Solution

She decided to construct a 99 percent confidence interval to answer the question. The value of p is given as .94 based on the sample class size of 9. The sample standard deviation is

$$s_p = \sqrt{(.94)(.06)/(9)} = \sqrt{.006267} = .0792$$

The t value corresponding to 99 percent confidence for the two-sided confidence interval is read from the student t table for 8 degrees of freedom and a two-tailed α of .01. The value is 3.355. According the formula, the confidence interval for the true percentage is

$$.94 + (3.355)(.0792) = 1.21$$
$$.94 - (3.355)(.0792) = .67$$

The upper confidence limit is calculated to be 1.21 or 121 percent, which is impossible. Because the largest possible percent is 100, or 1.00, the upper limit must be truncated at 1.00. Similarly, since the lowest possible value is 0.00, any negative percentages would have to be truncated at 0.

Confidence Intervals for Standard Deviations

Normal Populations

Although the calculated standard deviation gives the best estimate of the variability, it is still a point estimate. An interval estimate gives a better description of the location of the true population value for this statistic.

The formula for the interval estimate for the standard deviation from a normally distributed sample or population is calculated from the following formulas:

Upper limit:

$$s(n-1)/\sqrt{\chi^2_{\alpha/2}}$$

Lower limit:

$$s(n-1)/\sqrt{\chi^2_{1-\alpha/2}}$$

where

> s is the sample standard deviation
> n is the sample size

Chi-square (χ^2) is the table value for the χ^2 probability distribution reflecting the level of significance. (It is tabulated in Table F of the Appendix.) The degrees of freedom is equal to n − 1.

Example 5.8

The light bulbs used by Aft-Tech were inspected to determine how long they remained lit before burnout. The sample standard deviation of burnout times for a sample of 60 was 8.42 hours. Calculate the 95 percent confidence interval for the standard deviation of the burnout time.

Solution

Before the formulas for the upper and lower limits can be applied, the needed information must be identified. The sample standard deviation, s, is given as 8.42 hours. The sample size, n, is equal to 60. The use of the χ^2 probability distribution shown in Table F requires the identification of α. For 95 percent confidence, $\alpha = .05$. The values of χ^2 correspond to $\alpha/2$ and $1 - \alpha/2$. These are, respectively,

$$\alpha/2 = (.05)/(2) = .025$$
$$1 - \alpha/2 = 1 - .025 = .975$$

For n − 1 = 59 degrees of freedom, interpolation of the values in Table F gives values

$$\chi^2_{.025} = 81.89$$
$$\chi^2_{.975} = 39.89$$

The confidence limits are calculated from the formulas

Upper limit:

$$s\sqrt{(n-1)/\chi^2_{\alpha/2}} = 8.42\sqrt{59/39.68} = 10.27$$

Lower limit:

$$s\sqrt{(n-1)/\chi^2_{1-\alpha/2}} = 8.42\sqrt{59/81.53} = 6.09$$

The true value of the standard deviation can be said, with 95 percent confidence, to be between 6.09 and 10.27. The best single estimate is the point estimate, or the sample standard deviation of 8.42 hours.

Example 5.9

Aft-Tech determined, based on a sample of 11, that the machine producing the new toy Gizmos had an average defect rate of 6 percent. Calculate the 90 percent confidence interval for the standard deviation of the defect rate.

Solution

The sample standard deviation is 7.16 percent, or .0716. A 90 percent confidence interval for the sample standard deviation is calculated using the following:

$$s = .0716$$
$$n = 11$$
$$df = 10$$

Thus

$$\chi^2_{.050} = 3.94$$
$$\chi^2_{.95} = 18.31$$

The confidence limits are

$$.0716\sqrt{10/3.94} = .1141$$
$$.0716\sqrt{10/18.31} = .0529$$

Confidence Interval Recap

Table 5.1 shows the confidence intervals most frequently calculated for quality control applications.

Table 5.1

Statistic	Confidence Interval	df
Population mean Sigma unknown Large sample	$\bar{x} \pm \dfrac{z_{\alpha/s}s_x}{\sqrt{n}}$	
Population mean Sigma known Small sample	$\bar{x} \pm \dfrac{t_{\alpha/2}s_x}{\sqrt{n}}$	$n-1$
Standard deviation Upper	$s\sqrt{\dfrac{n-1}{\chi^2_{\alpha/2}}}$	$n-1$
Lower	$s\sqrt{\dfrac{n-1}{\chi^2_{1-\alpha/2}}}$	$n-1$
Proportion Large sample	$p \pm z_{\alpha/2}sp$	
Proportion Small sample	$p \pm t_{\alpha/2}sp$	$n-1$

The following problems are additional examples illustrating the calculation of confidence intervals. The problems are stated and solutions presented without commentary.

Example 5.10

Given

$$\bar{x} = 112 \quad s_x = 6.2 \quad n = 250 \quad \alpha = .05$$

Calculate the confidence interval.

Solution

$$112 \pm (1.96)(6.2) / \sqrt{250}$$

Upper = 112.769
Lower = 111.231

Example 5.11

Given

$$\bar{x} = 112 \quad s_x = 6.2 \quad n = 25 \quad \alpha = .05$$

Calculate the confidence interval.

Solution

$$112 \pm \frac{(2.064)(6.2)}{\sqrt{25}}$$

Upper = 114.559
Lower = 109.441

Example 5.12

Given

$$s = .137 \quad n = 17 \quad \alpha = .10$$

Calculate the confidence interval.

Solution

Upper:

$$(.137)\sqrt{16/7.96} = .194$$

Lower:

$$(.137)\sqrt{16/26.3} = .107$$

Example 5.13

Given

$$p = .36 \quad n = 150 \quad \alpha = .01$$

Calculate the confidence interval.

Solution

$$.36 \pm (2.575)\sqrt{\frac{(.36)(.64)}{150}}$$

Upper = .4609
Lower = .2591

Example 5.14

Given

$$p = .36 \quad n = 15 \quad \alpha = .01$$

Calculate the confidence interval.

Solution

$$.36 \pm (2.977)\sqrt{\frac{(.36)(.64)}{15}}$$

Upper = .729
Lower = −.009 = 0

Summary

This chapter has presented some introductory information about confidence intervals. Confidence, a measure of the probability of accepting what should be accepted, is important when any sampling activities are studied. Confidence intervals, which establish a range of values within which the true population value is very likely to reside, may be calculated for any parameter or statistic. Although the point estimate gives the best single estimate of the statistic, the confidence interval establishes a safety net that accounts for the variabilities expected in sampling.

In quality control applications, a 99.73 percent confidence interval is very often used. The 99.73 percent confidence corresponds to plus or minus 3σ, or three standard deviation limits.

Practice Problems

In problems 1 through 16 calculate the confidence interval.

1. $n = 120$ $\alpha = .05$ $\bar{x} = 12.7$ $s = 8$
2. $n = 750$ $\alpha = .01$ $\bar{x} = 412$ $s = 25$
3. $n = 1000$ $\alpha = .1$ $\bar{x} = 2950$ $s = 200$
4. $n = 6$ $\alpha = .1$ $\bar{x} = .083$ $s = .03$
5. $n = 14$ $\alpha = .05$ $\bar{x} = 111$ $s = .08$
6. $n = 22$ $\alpha = .01$ $\bar{x} = 18.243$ $s = .964$
7. $n = 12$ $\alpha = .05$ $p = .08$
8. $n = 20$ $\alpha = .1$ $p = .5$
9. $n = 16$ $\alpha = .1$ $p = .91$
10. $n = 175$ $\alpha = .01$ $p = .42$
11. $n = 100$ $\alpha = .05$ $p = .216$
12. $n = 1000$ $\alpha = .2$ $p = .875$
13. $n = 50$ $\alpha = .01$ $s = 1.64$
14. $n = 5$ $\alpha = .05$ $s = 16.4$
15. $n = 13$ $\alpha = .05$ $s = .164$
16. $n = 120$ $\alpha = .01$ $s = .068$

17. Measurements of 10 sample inside diameters on washers resulted in a sample mean of .43 centimeters and a sample standard deviation of .065 centimeters. Between what limits can one be 95 percent confident that the true inside diameter is found?

18. Measurements of 100 sample outside diameters on washers resulted in a sample mean of 4.21 centimeters and a sample standard deviation of .134 centimeters. Between what limits can one be 99 percent confident that the true outside diameter is found?

19. Measurements of 15 detergent bottles showed the average fill weight to be 15.86 ounces with a standard deviation of .55 ounces. Between what limits can one be 99 percent confident that the true fill weight is found?

20. Measurements of a sample of 28 tires indicated an average tread life of 42,500 miles with a standard deviation of 2,150 miles. Between what limits can it be said with 90 percent confidence that the true average tread life lies?

21. Inspection of 32 samples from a manufacturer indicated that 2 did not meet specifications. Between what limits can it be said with 95 percent confidence that the true proportion of defective material lies?

22. In the past month, 25 out of 150 samples have been rejected at final inspection. Between what limits can it be said with 99 percent confidence that the true proportion of rejected material lies?

23. The average number of defects produced last year was 16 per shift per day with a standard deviation of 2.5.
    (a) Determine 95 percent confidence limits for the mean and standard deviation based on a sample of 10 days.
    (b) Determine 95 percent confidence limits for the mean and standard deviation based on a sample of 100 days.
    (c) What is the difference between the two samples?

24. Five computer chips were produced. As part of the inspection procedure, the average time required to perform a very lengthy calculation was measured. The average time was 2.2 seconds with a standard deviation of .06 seconds. Within what limits can it be said, with 99 percent confidence, that the true processing time lies?

25. 500,000 potato chips were produced. As part of the inspection procedure, a sample of 200 were checked for percent of weight as oil. The sample check indicated an average of 28 percent. Calculate a 90 percent confidence interval for this statistic. What would have happened to this interval if a sample four times as big (800) had been taken? Can any conclusions be drawn about the appropriate sample size?

6 Tests of Hypotheses

Introduction

Decisions must often be made based on sample data. The statistical procedures that guide the decision making process are known as tests of hypotheses. Questions like the following often arise during the production of a product or delivery of a service:

- Is the new material as strong as the old?
- Is the new process really faster than the old?
- Does our product last longer than our competitor's?
- Do we provide faster service with the new method or the old?
- Has the process improvement reduced cycle time?
- Is there less variability with one tool than another?

In order to answer such questions, sample observations of the characteristic under consideration are made and descriptive statistics are calculated. These sample statistics are then analyzed, and the question is answered based on the results of the analysis.

Because the data used to answer the questions are sample data, there is always a chance that the answer will be wrong. If the sample is not truly representative of the population from which it was taken, the type I and type II errors discussed in Chapter 5 can occur. Thus, when a test of hypothesis is performed, it is essential that the confidence level — the probability that the statement is correct — be stated.

This chapter will first describe a test of hypothesis procedure that can be used to answer the initial question, whatever its nature. The procedure includes calculation of an appropriate test statistic from sample data. The balance of the chapter will illustrate the use of several of the more common test statistics. As in preceding chapters, *the information presented does not exhaustively cover the topic, but provides a foundation for the specific applications of statistical quality control.*

It must also be pointed out that the information that is being presented is a methodology that will assist the practicing quality professional in answering the types of questions posed above. Some liberties are taken with standard statistical methods in order to facilitate this objective.

Methodology

Stating the Hypothesis

When tests of hypotheses are to be used to answer questions, the first step is to state what is to be proved.

> The statement that is to be proved is known as the **null hypothesis**, or H_0.

For example, if a tire manufacturer wanted to determine whether a new tire design was the same as or better than the existing design, the null hypothesis might be expressed as follows:

> H_0: The new design is the same as the old design.

This hypothesis, as is the case with all of the hypotheses stated in this chapter, to be statistically correct, should be stated as:

> H_0: The new design is not significantly different from the old design.

The more convenient language is used because it relates better to the real question under investigation.

This statement is what the data analysis will attempt to prove or disprove. If the analysis shows that the statement is true, fine. But if the analysis indicates that the statement is not true, a fallback position is needed.

> A second hypothesis inconsistent with the null hypothesis is called the **alternative hypothesis,** or H_1.

In this example, if the data collected do not indicate that the two designs are the same, one design obviously must be better than the other. Three possible statements of the alternative hypothesis are

H_1: The new design is better than the old.
H_1: The old design is better than the new.
H_1: The two designs are not the same.

The conventional way to state hypotheses is in an algebraic format. The above null hypothesis and alternative hypotheses would be written, respectively, as:

H_0: old design = new design
H_1: old design < new design
H_1: old design > new design
H_1: old design ≠ new design

If the characteristic that determined the quality of the designs was tread life, then a more explicit statement of the hypotheses would be

H_0: average tread of old = average tread of new
H_1: average tread of old < average tread of new
H_1: average tread of old > average tread of new
H_1: average tread of old ≠ average tread of new

An even more concise way of stating this is as follows:

H_0: $X_{old} = X_{new}$
H_1: $X_{old} < X_{new}$
H_1: $X_{old} > X_{new}$
H_1: $X_{old} \neq X_{new}$

It is strongly recommended that the null hypothesis always be stated as an equality. Although this isn't necessary for statistical purposes, it does make later analysis much easier. The alternative hypothesis is then expressed either as the direction (less than or greater than) inequality or as a nondirectional inequality. The wording of the initial question determines the nature of the inequality used in the statement of the alternative hypothesis. A question involving "better than," "faster than," "stronger than," or similar terminology would require a directional inequality. The phrase "same as" or "not any different than" would imply a nondirectional inequality. The statement of the alternative hypothesis must be consistent with the observed sample data.

The nature of the inequality determines whether a test is one-tailed or two-tailed, a fact that is important in later steps of the analysis.

When the alternative hypothesis is stated as a directional inequality the procedure is called a **one-tailed** test of hypothesis.

A nondirectional inequality in the alternative hypothesis signifies a **two-tailed** test of hypothesis.

Specifying the Confidence Level

After both the null and the alternative hypotheses have been stated, the second step is to specify the confidence level. Usually the selection is arbitrary. However, there may be organizational guidelines that specify the confidence level. Common confidence levels are 90 percent, 95 percent, and 99 percent. A brief statement or an equation defining the confidence level in terms of α is usually sufficient; for example, the notation $\alpha = .05$ might appear after the hypothesis. This would designate 95 percent confidence.

Collecting Sample Data

The third step in testing hypotheses is the collection of sample data. After the null hypothesis has been identified — the equality of means, proportions, standard deviations, or whatever — the nature of the required data can be specified. These data must then be collected, and the appropriate sample descriptive statistics must be calculated.

Calculating Test Statistics

After the data have been collected and the sample test statistics have been calculated, the appropriate *test statistic* must be calculated. There are many test statistics that may be calculated, some of which will be illustrated later in this chapter. The specific test statistic used will depend on the nature of the null and alternative hypotheses.

Identifying Table Statistics

After the test statistic is calculated, the *table statistic* is determined. The nature of the alternative hypothesis, the sample size, and the specific statistic being tested will determine which of the standard distribution tables, such as the normal curve, student t, or chi-square, should be used. The examples throughout this chapter illustrate some of the table statistics used in testing hypotheses.

Decision Making

The most important aspect of the hypothesis testing procedure is the decision making step. If the above procedure has been followed step by step, and the null hypothesis has been stated as an equality, the following rule will govern all of the decisions, provided common sense is applied.

- If the absolute value of the test statistic is less than or equal to the table statistics, then there is not sufficient evidence to reject the null hypothesis. (In effect, but not in statistical preciseness, the null hypothesis is accepted as being true.)
- If the absolute value of the test statistic is greater than the table value, then there is sufficient evidence to reject the null hypothesis as being true. (By default this would imply that the alternative hypothesis must be true.)

Testing Various Hypotheses

The section illustrates some of the possible situations that might be encountered when hypotheses are tested. The test statistics will be recapitulated at the end of the chapter.

Population Mean, Sample Mean, and Population Standard Deviation Known

When the population mean, sample mean, and population standard deviation are known, the hypothesis that might be tested is as follows:

$$H_0: x = \mu$$
$$H_1: x \neq \mu \text{ or } H_1: x > \mu$$

The sample test statistic would be the following:

$$z = \frac{\bar{x} - \mu}{\sigma/\sqrt{n}}$$

where

  z is the test statistic
  $\bar{x}$ is the sample average
  μ is the population average
  σ is the population standard deviation
  n is the sample size

Example 6.1

Aft-Tech's tire division has long produced a steel belted radial tire that has an average 42,000 miles of tread life with a (population) standard deviation of 1,000 miles. A sample of 100 new tires has an average 44,500 miles of tread life. Can Aft-Tech say, with 95 percent confidence, that the new tire is (a) different from the old tire and (b) better than the old tire?

Solution

(a) In order to answer this question, the procedure for testing hypotheses must be followed. The null hypothesis is the equality

$$H_0 : \bar{x} = \mu$$

The alternative hypothesis contains the phrase "different from," so it is a nondirectional inequality

$$H_1 : \bar{x} \neq \mu$$

First the test statistic is calculated. The z value is

$$z = \frac{44,500 - 42,000}{1000/\sqrt{100}}$$

$$= \frac{2500}{1000/10}$$

$$= \frac{2500}{100} = 25$$

The table statistic must now be found. A nondirectional inequality dictates a two-tailed test, since the chance of making an error, α, is equally distributed both above and below the mean. Where there is a two-tailed test, the table statistic is derived using $\alpha/2$. For $\alpha/2 = .05/2 = .025$, the tail column in the normal curve table, Table C in the Appendix, gives $z = 1.96$.

The table statistic, 1.96, is less than the absolute value of the test statistic, 25, so the alternative hypothesis is selected. Thus it can be said, with 95 percent confidence, that the tread life of the sample tires is different from that of the tires that have been produced in the past.

(b) In order to determine whether the new tire is significantly better than the old tire, the alternative hypothesis must be expressed as a directional inequality:

$$H_1: x > \mu$$

The same test statistic is used whether the test is one-tailed or two. The difference lies in the table statistic. Because the test is now one-tailed, given the directional nature of the alternative hypothesis, the full value of α must be used to determine the table statistic, z.

For $\alpha = .05$ the table statistic is 1.645. This was again found in Table C. The absolute value of the test statistic, 25, is greater than 1.645, so the alternative hypothesis is again accepted. The sample tires last longer than the original tires.

Population Mean, Sample Mean, and Sample Standard Deviation Known

When the population mean, sample mean, and sample standard deviation are known, the hypothesis that might be tested is as follows:

$$H_0 : \bar{x} = \mu$$

$$H_1 : \bar{x} \neq \mu \ \text{ or } \ H_1 : x > \mu$$

The sample test statistic would be the following:

$$t = \frac{\bar{x} - \mu}{s/\sqrt{n}}$$

where

>   t is the test statistic
> $\bar{x}$ is the sample average
> μ is the population average
>   n is the sample size
>   s is the sample standard deviation

The degrees of freedom, df, is equal to the sample size (n) minus 1 (n − 1).

Example 6.2

The blade manufacturing division of Aft-Tech has determined that, in the past, the silicon coating for the X-TR-A has had an average weight of 16.4 grams. A sample of seven blades taken at random from the completed work has an average silicon coating weight of 16.94 grams. The sample standard deviation is 1.1 grams. Is the sample blade significantly different, at the .1 level of significance, from the blade that has always been produced by Aft-Tech?

Solution

In order to answer this question, the hypotheses must be stated. The null hypothesis, as always, is an equality.

$$H_0 : \bar{x} = \mu$$

The alternative hypothesis contains the phrase "different from," so it is a nondirectional inequality

$$H_1 : \bar{x} \neq \mu$$

The test statistic is calculated

$$t = \frac{16.94 - 16.4}{1.1/\sqrt{7}} = (.54/.42) = 1.29$$

The table statistic for a two-tailed test of hypothesis with $\alpha = .10$ and df = 6 is found from the student t table, Table E in the Appendix, to be 1.943. Since the test value, 1.29, is less than the table value, 1.943, it is determined that there is not sufficient evidence at the .1 level to reject the null hypothesis. There is no significant difference between Aft-Tech's former blade and its present blade. (In effect the null hypothesis is accepted, although in statistics we never really accept the null hypothesis.)

Example 6.3

The light bulb division of Aft-Tech produces bulbs that have an average life of 867 hours. A new process produces bulbs that last 899 hours with a sample standard deviation of 16 hours, based on a sample of 8. Can Aft-Tech say, with 90 percent confidence, that the new process is significantly different from the old?

Solution

The hypotheses are

$$H_0 : \bar{x} = \mu$$

$$H_1 : \bar{x} \neq \mu$$

The test statistic is calculated

$$t = \frac{899 - 867}{16/\sqrt{8}} = 32/5.66 = 5.657$$

The table value of t corresponding to a two-tailed test with $\alpha = .1$ and $n - 1 = 7$ degrees of freedom is 1.895. The fact that the test value is larger than the table value indicates that there is enough data to reject the null hypothesis. Thus there is a difference between the sample mean, representing the new, and the population mean, representing the old.

Two Sample Means Equal, Sample Standard Deviations Known

When the means and sample standard deviations of two samples are known, the hypothesis that might be tested is as follows:

$$H_0 : \overline{x}_a = \overline{x}_b$$

$$H_1 : \overline{x}_a \neq \overline{x}_b \text{ or } H_1 : \overline{x}_a > \overline{x}_b$$

The sample test statistic would be the following:

$$t = \frac{\overline{x}_a - \overline{x}_b}{\sqrt{\dfrac{1}{n_1} + \dfrac{1}{n_2}} \sqrt{\dfrac{(n_1 - 1)(s_1^2) + (n_2 - 1)(s_2^2)}{n_1 + n_2 - 2}}}$$

where

$\overline{x}_a$ is the first sample average
$\overline{x}_b$ is the second sample average
n_1 is the first sample size
n_2 is the second sample size
s_1 is the sample standard deviation of the first sample
s_2 is the sample standard deviation of the second sample

The degrees of freedom is equal to the sum of the two sample sizes minus 2.

Example 6.4

Aft-Tech has developed a new silicon chip that is supposed to perform calculations in .0006 milliseconds. Archrival ForeTech claims to have developed a chip that will perform the same calculations in .0005 milliseconds. Ten of Aft-Tech's chips are tested and the mean performance time is .0006 milliseconds with a sample standard deviation of .00009 milliseconds. Fifteen of ForeTech's chips have the claimed average performance time of .0005 milliseconds with a sample standard deviation of .00016 milliseconds. Although .0005 is obviously less than .0006, can it be said with 99 percent confidence that the Aft-Tech chip is slower than the ForeTech chip? *Although one sample has a lower average than the other, there may not be a significant difference between the values.*

Solution

Only the test of hypothesis procedure will tell for sure. The hypotheses in this instance are

$$H_0 : \bar{x}_a = \bar{x}_f$$

$$H_1 : \bar{x}_a > \bar{x}_f$$

The test statistic is calculated from the sample data

$$t = \frac{.0005 - .0006}{\sqrt{\frac{1}{10} + \frac{1}{15}} \sqrt{\frac{(9)(.000009)^2 + (14)(.00016)^2}{10 + 15 - 2}}}$$

$$= \frac{-.0001}{\sqrt{.1667} \sqrt{.000000019}}$$

$$= -.0001/.000055936 = -1.788$$

The alternative hypothesis in this instance is a directional inequality, which implies a one-tailed test. The α value for this case is .01, with df = 23. From the t table the table statistic is 2.500. Since the absolute value of the test statistic, 1.788, is less than the table value of 2.500, there is not enough evidence to reject the null hypothesis. Thus it can be said that the ForeTech chip is equal in speed to the Aft-Tech chip. Although there is an observed difference, there is no statistically significant difference between the two values based on the sample data collected.

Example 6.5

The students in two first-grade classes in the local school system were weighed. One class of 20 students had an average weight of 55.4 pounds with a standard deviation of 6 pounds. The second class of 15 students had an average weight of 59.1 pounds with a sample standard deviation of 4.3 pounds. At the .05 level, is there a significant difference between the weights of the classes?

Solution

The hypotheses for this example are

$$H_0 : \bar{x}_1 = \bar{x}_2$$

$$H_1 : \bar{x}_1 \neq \bar{x}_2$$

The test statistic is calculated

$$t = \frac{56.4 - 59.1}{\sqrt{\dfrac{1}{20} + \dfrac{1}{15}}\sqrt{\dfrac{(19)(6)^2 + (14)(43)^2}{20 + 15 - 2}}} = \frac{-3.7}{\sqrt{11.7}\sqrt{28.572}}$$

$$= \frac{-3.7}{(.342)(5.345)} = -\frac{3.7}{1.828}$$

$$= -2.0266$$

The α value has been specified as .05. For this two-tailed test with $n_1 + n_2 - 2 = 33$ df, the table value is 2.0357. The absolute value of the test statistic, 2.0266, is less than the value read from the table, 2.0357, so the null hypothesis cannot be rejected. There is not a statistically significant difference between class weights based on the sample data collected.

Sample Proportion Equal to Population Proportion, Population Proportion Known

When the sample proportion is equal to the population proportion and the population proportion is known, the test statistic is

$$z = (p - \pi)\big/\sqrt{(\pi)(1 - \pi)/n}$$

where

       z is the test statistic
       p is the sample proportion or percentage
       π is the population proportion or percentage
       n is the sample size

Example 6.6

Aft-Tech knows that in the past a certain supplier has always supplied material that was 6 percent defective. A recent sample of 150 units was 3.5 percent defective. Can Aft-Tech say, with 90 percent confidence, that the quality of the material has improved?

Solution

The hypothesis is stated as follows:

$$H_0: p = \pi$$
$$H_1: p < \pi$$

The equality, as always, is the null hypothesis. Even if the goal is to prove the alternative hypothesis, the null hypothesis should always be stated as the equality. The test statistic can be calculated from the sample data collected:

$$z = \frac{.035 - .060}{\sqrt{(.06)(.94)/150}} = \frac{-.025}{\sqrt{.00376}}$$

$$= \frac{-.025}{.01939} = -1.289$$

An α value of .1 has been specified. For $\alpha = .1$, the table value for the one-tailed test of hypothesis is 1.282. Because the absolute value of the test value is greater than the test value, the null hypothesis cannot be accepted and is thus rejected. There are statistical grounds for the assertion that the quality has improved.

Two Sample Proportions Equal

When two sample proportions are equal, the test statistic is

$$t = \frac{p_1 - p_2}{\sqrt{\dfrac{p_1(1 - p_1)}{n_1} + \dfrac{p_2(1 - p_2)}{n_2}}}$$

where

    t is the test statistic
    p_1 is the proportion in the first sample
    p_2 is the proportion in the second sample
    n_1 is the size of the first sample
    n_2 is the size of the second sample

The degrees of freedom is equal to the sum of the two sample sizes minus 2.

Example 6.7

Aft-Tech has two suppliers under consideration to serve as vendor for a particular product. Sample lots from each supplier were inspected. The first sample of 16, taken from material provided by vendor 1, had 2 defects. Material from vendor 2 had 1 defect in the 10 units provided. Can Aft-Tech say, with 99 percent confidence, that there is a significant difference between the proportions of defective material provided by the two vendors?

Solution

The hypotheses are stated

$$H_0: p_1 = p_2$$
$$H_1: p_1 = p_2$$

In order to determine the test statistic, sample proportions must be calculated for the sample data.

$$p_1 = 2/16 = .125 \qquad p_2 = 1/10 = .1$$

The value of the test statistic is

$$t = \frac{.125 - .100}{\sqrt{\dfrac{(.125)(.875)}{16} + \dfrac{(.1)(.9)}{10}}} = \frac{.025}{\sqrt{.01583}}$$

$$= \frac{.025}{.12584} = .198$$

For $\alpha = .01$ and df $= 16 + 10 - 2 = 24$, the table value (from Table E) corresponding to the two-tailed test is 2.797. The fact that the calculated test value is smaller than the table value implies that there is not sufficient data to reject the null hypothesis. Thus it can be said that there is no statistically significant difference between the sample proportions of defective items observed.

Example 6.8

A certain college professor taught two sections of the same course. The first class of 20 students had an average grade of 72 on the final exam. The second class of 7 students had an average grade of 81 on the same test. Can this professor say, with 95 percent confidence, that the small class did better than the large class?

Solution

The hypotheses are stated

$$H_0: p_1 = p_2$$
$$H_1: p_1 < p_2$$

The test statistic is calculated from the sample data

$$t = \frac{.72 - .81}{\sqrt{\dfrac{(.72)(.28)}{20} + \dfrac{(.81)(.19)}{7}}} = \frac{-.09}{\sqrt{.0320}}$$

$$= \frac{-.09}{.1790} = .502$$

For $\alpha = .05$ and df = 25, the table value (from Table E) corresponding to the one-tailed test is 1.708. The absolute value of the test statistic, .502, is less than the table value, 1.708, so the null hypothesis cannot be rejected. There is no significant difference, at the .05 level, between the classes' scores on the final exam.

Sample Standard Deviation Equals Population Standard Deviation

When the sample standard deviation equals the population standard deviation, the test statistic is

$$\chi^2 = (n - 1)(s^2)/\sigma^2$$

where

χ^2 reflects the confidence
n is the sample size
s is the sample standard deviation
σ is the population standard deviation

The degrees of freedom is equal to the sample size minus 1. This is a two-tailed test of hypothesis. The null hypothesis should always be expressed as the equality

$$H_0: s = \sigma$$

The alternative hypothesis is then expressed as the nondirectional inequality

$$H_1: s \neq \sigma$$

The table value of the χ^2 statistic corresponds to $1 - \alpha/2$ and $n - 1$ degrees of freedom.

Example 6.9

Aft-Tech has just purchased some electronic components. A sample inspection of 15 of these revealed a sample performance measure of 98 with a sample standard deviation of 12. Traditionally this component has had an average performance of 100 with a standard deviation of 14.2. Is this measure significantly better, at the .01 level, than that of the product traditionally received by Aft-Tech?

Solution

Before testing for the equality of the averages, it was decided that it would be necessary to make sure that the standard deviation had not changed. The null hypothesis is

$$H_0: s = \sigma$$

The alternative hypothesis for this two-tailed test is

$$H_1: s = \sigma$$

The test statistic is calculated

$$\chi^2 = \frac{(14)(12)^2}{(14.2)^2} = 9.998$$

At the $\alpha = .01$ level, the key value of $1 - \alpha/2$ is $1 - .01/2 = 1 - .005 = .995$. The χ^2 table value, found in Table F of the Appendix, is 31.32. Because the calculated value of the test statistic is smaller than the table value, the null hypothesis cannot be rejected. There is, at the .01 level (or with 99 percent confidence) no significant difference between the sample standard deviation and the population standard deviation.

In order to completely answer the question, "Is there a significant difference between the sample and population?" it is necessary to show that the sample and the population not only have the same mean, but also the same variability, as measured by the standard deviation. Thus a second test of hypothesis must be performed on the sample and population means. The hypotheses for this test are

$$H_0 : \bar{x} = \mu$$

$$H_1 : \bar{x} < \mu$$

The sample test statistic is

$$z = \frac{(100 - 90)}{(12/\sqrt{15})} = 2/3.1 = .65$$

The table value of z corresponding to 99 percent confidence and a one-tailed test of hypothesis is 2.642. The calculated value is less than the table value, so the null hypothesis cannot be rejected. Now it can be said that there is no significant difference between the component.

Two Sample Standard Deviations Equal

When two sample standard deviations are equal, the test statistic is the ratio of the sample variances.

$$F = s_1^2 / s_2^2$$

where

F is the test statistic
s_1 is the sample standard deviation of the first sample
s_2 is the sample standard deviation of the second sample

The null hypothesis is

$$H_0: s_1 = s_2$$

The alternative hypothesis is

$$H_1: s_1 \neq s_2$$

The table statistic comes from the F distribution. This is provided in Table G of the Appendix.

Because of the way the F distribution has been tabulated, *a special stipulation for this test of hypothesis is that $s_1 \geq s_2$*. The degrees of freedom for each sample is equal to the sample size minus 1.

Example 6.10

Two machines are used by Aft-Tech to produce a certain product. Products from machine 1 have a sample standard deviation of .086 centimeters and a sample average of 4.81 centimeters, based on inspection of 11 parts. A sample of 16 products from machine 2 has a sample standard deviation of .069 centimeters and a sample average of 4.89 centimeters. (a) Can it be said, with 95 percent confidence, that the samples have the same variability? (b) Is there a significant difference, at the .05 level, between the sample means?

Solution

(a) For the standard deviation, the hypotheses are

$$H_0: s_1 = s_2$$
$$H_1: s_1 \neq s_2$$

The test statistic is

$$F = (.086)^2/(.069)^2 = 1.55$$

The table value corresponding to a .05 level of significance and $n_1 - 1 = 11 - 1 = 10$ degrees of freedom for the numerator and $n_2 - 1 = 16 - 1 = 15$ df for the denominator is found from Table G to be 2.54. Because the sample test value of 1.55 is less than the table value, there is a 95 percent chance that there is no significant difference between sample standard deviations.

(b) For the means, the hypotheses are

$$H_0 : \bar{x}_1 = \bar{x}_2$$

$$H_1 : \bar{x}_1 \neq \bar{x}_2$$

The sample test statistic is calculated from the formula

$$t = \frac{4.81 - 4.89}{\sqrt{\frac{1}{11} + \frac{1}{16}}\sqrt{\frac{(10)(.086)^2 + (15)(.069)^2}{11 + 16 - 2}}}$$

$$= -2.678$$

The table value corresponding to a two-tailed α of .05 and $n_1 + n_2 - 2 = 25$ degrees of freedom is 2.060. Since the absolute value of the test statistic is larger than the table value, the null hypothesis cannot be accepted. It cannot be said, with 95 percent confidence, that the sample averages are the same. Although the dispersion is not significantly different, the two machines are centered at significantly different means.

Test Statistic Recap

Table 6.1 lists null hypotheses and the corresponding test statistics. The table values for the tests enumerated in the table depend on the confidence level and whether the alternative hypothesis specified is a one-tailed or two-tailed test.

Following are several additional example problems. After the problem statement a solution is presented without comment. The reader is left the task of answering the question, based on the statistics that have been calculated.

Example 6.11

Time standards for Aft-Tech show that a certain inspection task should take .43 hours. As part of an audit, 14 cycles of this inspection were timed. The average cycle time was .48 hour with a sample standard deviation of .092 hours. Can Aft-Tech's standards analyst conclude, with 95 percent confidence, that this inspection is taking longer than it should?

Table 6.1

Null Hypothesis	Special Conditions	Test Statistic	df
$H_0 : \bar{x} = \mu$	σ known	$z = \dfrac{x - \mu}{\sigma/\sqrt{n}}$	
$H_0 : \bar{x} = \mu$	s known	$t = \dfrac{x - \mu}{s/\sqrt{n}}$	$n - 1$
$H_0 : \bar{x}_1 = \bar{x}$		$t = \dfrac{x_1 - x_2}{\sqrt{\dfrac{1}{n_1} + \dfrac{1}{n_2}}\sqrt{\dfrac{(n_1-1)(s_1^2) + (n_2 - 1)}{n_1 + n_2 - 2}}}$	$n_1 + n_2 - 2$
$H_0 : p = :$		$z = \dfrac{p - \pi}{\sqrt{(\pi)(1-\pi)/}}$	
$H_0 : p_1 = p$		$t = \dfrac{p_1 - p_2}{\sqrt{\dfrac{(p_1)(1-p_1)}{n_1} + \dfrac{(p_2)(1-}{n_2}}}$	$n_1 + n_2 - 2$
$H_0 : s = ($		$\chi^2 = \dfrac{(n-1)(s')}{\sigma^2}$	$n - 1$
$H_0 : s_1 = s$	$s_1 \geq s_2$	$F = \dfrac{s_1^2}{s_2^2}$	$n_1 - 1$ and $n_2 - 1$

Solution

$$H_0 : \bar{x} = \mu$$

$$H_1 : \bar{x} > \mu$$

$$t_{test} = \frac{.48 - .43}{.092/\sqrt{14}} = 2.033$$

$$t_{table} = 1.771$$

Example 6.12

Aft-Tech's company airplane has maintained a 94 percent on-time record during the past year. A recent sampling of 20 flights showed that 17 were on time. Can it be said, with 90 percent confidence, that the on-time performance of the company plane has changed?

Solution

$$H_0 : p = \pi$$

$$H_1 : p = \pi$$

$$p = 17/20 = .85$$

$$z_{test} = \frac{.85 - .94}{\sqrt{(.94)(.06)/20}} = -1.695$$

$$z_{table} = 1.645$$

Example 6.13

A certain department at one of Aft-Tech's manufacturing plants has averaged, during the past 13 weeks, 7 percent defective material. During the previous 13-week period, the average was 4 percent defective. Can it be said, with 99 percent confidence, that (a) the performance of the department has deteriorated and (b) the variability of the department during the two quarters was the same?

Solution

(a)

$$H_0: p_1 = p_2$$
$$H_1: p_1 > p_2$$
$$t_{test} = .0825$$
$$t_{table} = 2.492$$

(b)

$$H_0 : s_1 = s_2$$
$$H_1 : s_1 = s_2$$
$$s_1 = \sqrt{\frac{(.07)(.93)}{13}} = .0708$$
$$s_2 = \sqrt{\frac{(.96)(.04)}{13}} = .0543$$
$$F_{test} = (.0708)^2 / (.0543)^2 = 1.697$$
$$F_{table} = 4.16$$

Summary

Tests of hypotheses are useful guides for decision making. Based on sample data relating to a process, a product, or material, inferences can be drawn about the nature of the population represented. The inferences, which have an element of risk indicated by the accompanying statement of confidence, serve as a guide to — not a substitute for — management decision making.

When a null hypothesis is accepted, all that is implied is that there is no statistical reason to believe that the statement is not true. There is no statistically significant difference between the values being tested. Nonacceptance of a null hypothesis means that there is enough statistical reliability in the sample data to indicate that the original statement is not true. It is important that the tester's common sense be applied to the statement of the hypotheses. The hypotheses must agree with the observed data.

Practice Problems

In problems 1 through 7 test the hypotheses indicated.

1. $\alpha = .01$ $H_0: p = \pi$ $H_1: p \neq \pi$ $\pi = .42$ $\pi = .48$ $n = 100$
2. $\alpha = .05$ $H_0: \bar{x}_1 = \bar{x}_2$ $H_1: \bar{x}_1 > \bar{x}_2$ $n_1 = 8$ $n_2 = 7$
   $\bar{x}_1 = 6.98$ $\bar{x}_2 = 5.98$ $s_1 = 1.04$ $s_2 = 1.66$
3. $\alpha = .1$ $H_0: \bar{x} = \mu$ $H_1: \bar{x} = \mu$ $\bar{x} = 98.6$ $\mu = 99.9$ $\sigma = 2.14$
   $n = 13$
4. $\alpha = .05$ $H_0: s = \sigma$ $H_1: s = \sigma$ $\sigma = 41$ $s = 52$ $n = 16$
5. $\alpha = .1$ $H_0: p_1 = p_2$ $H_1: p_1 > p_2$ $p_1 = .19$ $p_2 = .12$
   $n_1 = 22$ $n_2 = 19$
6. $\alpha = .01$ $H_0: s_1 = s_2$ $H_1: s_1 = s_2$ $s_1 = .065$ $s_2 = .085$
   $n_1 = 6$ $n_2 = 12$
7. $\alpha = .05$ $H_0: \bar{x} = \mu$ $H_1: \bar{x} < \mu$ $\mu = .919$ $s = .188$ $n = 20$

8. Aft-Tech's corporate hotel has averaged a 78 percent occupancy rate. For the past seven days, the occupancy rate has been 83 percent. Can it be said, at the 95 percent confidence level, that occupancy has increased?

9. Ralph Zenith, noted consumer advocate, has decided to test the claims that tire manufacturer A has been making regarding manufacturer B. Ralph tested 18 of A's tires and found an average tread life of 41,400 miles with a standard deviation of 4,250 miles. Manufacturer B's 20 tires lasted an average 39,600 miles, with a standard deviation of 3,960 miles. At the .01 level, is there a significant difference in tire lives?

10. A large department store chain has averaged 16 complaints per day on its hotline, with a standard deviation of 4 calls per day. The average for a recent sample of 14 days was 13 complaints. At the .01 level can it be said the complaint rate has decreased?

11. One class of 20 students at State Tech had an average grade of 78 with a standard deviation of 4 in the required quality control class. Historically, students in this class had an average grade of 74. Is this class significantly different at the .05 level, from past classes?

12. Aft-Tech's packaging inspector has discovered that on the average, plant 4 fills its packages to 96 percent of capacity, based on a sample of 20, whereas plant 6 fills its packages to only 88 percent of capacity, based on a sample of 10 packages. Can the inspector say, with 99

13. Is there a significant difference in the variability of the packaging levels as reported in problem 12?

14. Aft-Tech measured employees at the new South American operation to determine how they compared in height with the North American employees. The company discovered that, based on a sample of 25 employees, the average height of the South American employees was 176.53 cm with a standard deviation of 12.17 cm. The North American employees' average height, corporate wide, was 183.41 cm with a standard deviation of 9.71 cm. At the .05 level, is there a significant difference in (a) heights and (b) variability?

15. Two samples were selected from Aft-Tech's screw machine operation. The first sample, produced on the day shift, had a mean of .87 and a standard deviation of .91, based on a sample of 12 screws. The second sample, produced on the night shift, had an average of .96 with a standard deviation of 1.09, based on a sample of 10 screws. (a) At the .05 level, is the night shift sample mean larger than the mean for the day shift? (b) At the .05 level, is the variability in the night shift different than in the day shift?

16. Aft-Tech's lawn control service has traditionally served an average of 19 customers a day. A five-day sample showed Aft-Tech serving an average of 14 customers with a standard deviation of 6.2 customers. At the .1 level, can it be said the number of customers served has decreased?

17. State Tech wanted to determine if there was a difference in the dropout rates between the QA majors and the EE majors. Twenty students in each department were tracked, and it was noted that four QA students and seven EE student dropped out. At the .05 level can the QA dropout rate said to be lower than the EE dropout rate?

18. Sampling 12 computer stores, Aft-Tech discovered that in a month's time an average of 15 of its personal computers were sold by each store, with a standard deviation of 2.5. A similar study of the chief competitor indicated a monthly average of 12 computers per store, with a standard deviation of 2.9. Can it be said, with 99 percent confidence, that the Aft-Tech personal computer outsells the competition?

7 Control Charts

Introduction

Often the term "quality product" is taken to mean a product that performs at or near perfection. A more practical definition of quality focuses on consistency. A quality product is a product that consistently performs in the expected manner. In most production applications, consistency, producing the product virtually the same way every time, is more important than occasionally producing a perfect product. Consumers develop product expectations based on past performance, and they often change their buying habits when anything happens to change this perception. Potential customers often demand evidence that a product is produced uniformly before they start to use the product.

Control charts are the statistical tool that permit an organization to demonstrate and monitor the consistency of its process or product. They provide a record of current performance and deviations from past performance. Control charts are based on the statistical principles discussed in previous chapters.

There are several types of control charts, each of which can be used to chart a number of different characteristics. Variables charts are used when characteristics can be objectively and quantitatively measured; attribute charts are used when characteristics can be counted. Cumulative sum, or cusum, charts are a special type of control chart based on sequential sampling.

This chapter will introduce the methodology of control charts, specifically the variables charts for average, range, individuals, and standard deviations, and the attribute charts for percentage defective and defects per unit. There will also be a brief introduction to the cusum chart.

Variability

Although consistent perfection is a common goal, few ever achieve it. Perfection implies that everything comes out exactly the same every time, and the average person cannot even sign his or her own name the same way two times in a row. As a matter of fact, identical signatures are often a sign of forgery.

No matter how many products are produced, whether they be steel bars, plastic pellets, cars, pens, television, or textbooks, no two are ever exactly the same. Rarely does anyone purchase a car that performs entirely as it should. Sometimes even the pen that's supposed to write the first time every time doesn't. Random, or chance variation is present all the time. No two cars perform exactly the same way and no two pens write the same way because both are subject to random variation, which is uncontrollable.

> **Random variation** is the variation that occurs as a result of natural causes. It is normal for all processes to have random variation.

Example 7.1

After archer A1 shot a round of arrows and then pulled them out, the target looked like Figure 7.1. Why were there multiple holes in the target instead of just one in the absolute center?

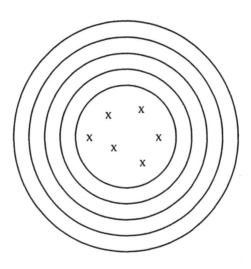

Figure 7.1   Larget Showing Random or Chance Variation

Solution

When archer A1 shoots an arrow at a target, he hopes to hit the bull's eye every time. But not being a mythical archer of the Robin Hood or William Tell class, he can reasonably expect his arrows to land on different parts of the bull's eye, as a result of random variation. Many chance causes, all of which were beyond the control of the archer, probably contributed to the scattered pattern. For example, the bowstring was probably pulled a little bit differently each time an arrow was shot, the breeze undoubtedly changed imperceptibly between shots, and there were probably some subtle differences between each of the arrows.

In general, random or chance variation exhibits the following characteristics:

- It has many individual causes.
- Any one cause results in only a small amount of variation.
- Many causes act simultaneously so that the total amount of variation is significant.
- It cannot be eliminated.

Not all variation is randomly caused. Sometimes the target is missed for a very explicit reason: the bowstring broke just as the arrow was being released, the wind suddenly gusted, or the archer just didn't aim well. Whatever the reason, if there is a definite reason, the variation is considered assignable and due to special causes.

Assignable variation is variation that has a definite cause or set of special causes.

Assignable variation is not present all the time. When it is present, though, it often causes undesirable results. Dr. Deming has estimated that assignable variation due to special causes is responsible for approximately 15 percent of all quality problems.

Example 7.2

The inspector in one of Aft-Tech's machine shops noticed the pattern of inspection readings shown in Figure 7.2 for one of the punch press operations. The readings show that the process is now producing parts that fall well out of the specification limits. The pattern indicates that all of the

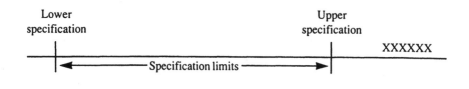

Figure 7.2   Relationship between Specifications and Measurements

inspected parts have experienced a marked shift to the right, or the high side. Is this random or assignable variation?

Solution

Some form or forms of assignable variation have definitely entered the manufacturing process. Perhaps the cutting tool in the punch press die has worn slightly and is therefore producing parts that are not meeting requirements. Other potential causes of this assignable variation are broken tools, poor quality material, wrong material, and inspection errors.

Assignable variation has many potential causes. Once assignable variation is noted, its cause or causes should be identified and, if possible, eliminated if the variation is harmful to the product or process. Some of the characteristics of assignable variation are summarized here.

- Assignable variation has several causes.
- Any one cause may or may not be responsible for a large amount of variation.
- Causes can often be easily located or identified.
- Causes can often be easily located or identified by the people doing the work.
- Depending on the specific process and how it is affected, it may or may not be economically feasible to try to eliminate the cause or causes.

All processes, whether controlled by people or by machines have some variation. It is important that the quality analyst be able to recognize abnormal variation when it occurs in order to take the corrective actions necessary to maintain quality standards.

Control Charts for Variables

Control charts are graphical tools that show variation within a process and allowable or expected variation of a process over a period of time.

Control charts require that sample values of a particular variable be recorded periodically. The average of these sample values will form, when a large enough sample is taken, a distribution that will always approach the normal distribution. The standard deviation of these averages, called the standard error of the mean, is smaller by a factor of $1/\sqrt{n}$ than the standard deviation of the individual values. As sampling continues, any significant change in the sample average will be reflected by a noticeable change in the shape of the distribution, indicating that assignable variation is present in a process. Thus unexpected changes are noticed quickly when control charts monitor the performance of the process.

Control limits are values between which a certain percentage of measurements are normally expected to fall.

They are, in effect, confidence intervals for the measurements. The control limits for variables commonly used in industry are based on parameters determined empirically by Shewhart. These limits are usually called three sigma limits.

Three sigma limits are the control limits between which it is normally expected that 99.73 percent of the values will be found.

Sometimes two sigma limits are used, particularly in the health care field and aerospace industry. With two sigma limits, 95.5 percent of the possible values are normally expected to fall between the control limits.

Control limits are not difficult to calculate and are quite easy to use once they have been calculated. But they are effective only when management permits them to be used as they were originally intended to be used — as identifiers of assignable or special causes of variation.

Construction and Use of Control Charts

Control charts show the control limits for any process. A high reject rate, introduction of a new manufacturing process, customer complaints, or just normal operating procedures may indicate that control charts should be maintained for a particular process.

Control charts for variables may be kept for individuals, averages, ranges, and standard deviations. Individual control charts show specifically how the process is performing in relation to expectations and to quality specifications for the process. Averages, because of the special relationship developed on the basis of the central limit theorem, show variation much more quickly than do the individuals. Ranges show the variability within the process. For large sample sizes, usually 10 or greater, a sample standard deviation minimizes the impact on variation that a single outlier might have. A variability that is either too large or too small indicates that assignable causes of variation exist.

Control charts for attributes show how characteristics, such as percentage defective or defects per unit, normally vary over time. Significant changes in these characteristics also indicate that special causes of variation have entered the process.

Before a control chart can actually be constructed, frequency of sampling and sample size must be determined. The main determinant for frequency of sampling is the nature of the process. Some processes have to be sampled continuously, whereas others may need sampling only once a day. The specification of frequency of sampling is left to the judgment of the quality control analyst. Sample size may be determined scientifically; however, it is usually easier and almost as accurate to select a sample size of between 2 and 10 units at a time. These characteristics are usually specified in an in-process inspection instruction sheet, such as the one shown in Figure 7.3. A specification might read, "Inspect 7 parts each hour." Once the frequency of sampling and the sample size have been determined, the control chart itself can be developed.

Following is the method for developing and using a Shewhart control chart for three sigma limits for sample averages and sample ranges.

1. Specify the part, product, or characteristic to be charted. There are so many potential candidates for charting that one cannot reasonably expect to chart all of them. Only those deemed to be most important should be charted. (Note: See a text such as *Quality Improvement Using Statistical Process Control* to read about the methods used to identify the important process characteristics that should be charted.)
2. In order to establish precontrol limits, select a sample size of between 2 and 10. This sample size *must* be kept constant during the analysis.
3. Conduct a base period (or precontrol) analysis to determine the limits. Usually, a minimum of 50 measurements should be made of the variables under study — for example, 10 samples of size 5. The important thing to remember is that the samples of data collected

DATE _____ QC# _____
PRINT ISSUE NO. _____
WRITTEN BY _____
APPROVED BY _____

QUALITY CONTROL
PROCESS INSPECTION INSTRUCTIONS

PART NO. _____
PART NAME _____
DEPT _____
OPER _____

SHEET ____ OF ____ SHEETS

OTHER ITEMS TO BE CHECKED

SAMPLE	CHARACTERISTIC

TYPE OF
CHART
☐ X̄ & R - PLOT AVG AND RANGE
☐ X̄ & R - PLOT INDIVIDUALS AND RANGE
☐ np - PLOT NO. OF DEFECTS FOUND
☐ OTHER -

UNIT	SAMPLE SIZE	TIME INTERVAL AFTER SET UP OR RESET.
CHARTERED CHARACTERISTIC		____ HOUR LATER EVERY ____ HOUR THEREAFTER

SAMPLES
TO BE
TAKEN
☐ CONSECUTIVE PEICES FROM OPERATION
☐ AT RANDOM FROM WORK COMPLETED SINCE LAST INSPECTION

CONTROL LIMITS
X̄
R

GENERAL INSTRUCTIONS

FOLLOW TIME INTERVAL CLOSELY
RECORD INFORMATION NEATLY
NOTE INFORMATION ON CHART REGARDING MACHINE RESETS.
SUPERVISOR NOTIFIED, ETC.
DO NOT RECORD SET-UP INSPECTION UNLESS SET-UP IS APPROVED.
CHARTS MUST BE TURNED IN TO QUALITY CONTROL AT COMPLETION
OF CHART, END OF MONTH, OR END OF JOB.

1	QC CHART FILE	1	M.I. STAFF
2	INSP. DEPARTMENT	2	QC ENGR FILE

Figure 7.3   Industrial In-Process Inspection Instruction Sheet

need to be representative of the process characteristics being sampled. These values should be recorded on a form, as shown in Figure 7.4. (Some control chart forms incorporate data collection areas on the chart.)

4. Based on these initial sample data, make two preliminary calculations. First, calculate the sample average for each sample, $\bar{x}$. Second, compute the sample range, R, the largest value minus the smallest value within the sample.

5. Determine the average, $\bar{\bar{x}}$, of the sample averages. This represents the population or process average.

6. Determine the average, $\bar{R}$, of the sample ranges. It is a measure or estimate of variability of the samples.

7. Calculate the standard deviation of the individuals. Shewhart's approximation, called σ', can be used to make the task less tedious. Based on his empirical data, Shewhart defined σ' as

$$\sigma' = \bar{R}/d_2$$

where d_2 is a Shewhart control chart constant found in Table H in the Appendix, whose value depends on the sample size, n.

8. Calculate the control limits using the following sets of equations. The control limits for the sample averages are those within which 99.73 percent, or virtually all, of the sample averages are expected to fall when only random or chance variation is present. These limits are given by the equations:

$$UCL_{\bar{x}} = \bar{\bar{x}} + A_2\bar{R}$$
$$LCL_{\bar{x}} = \bar{\bar{x}} - A_2\bar{R}$$

where

$UCL_{\bar{x}}$ is the upper control limit for averages
$LCL_{\bar{x}}$ is the lower control limit for averages
$\bar{\bar{x}}$ is the process average
A_2 is a Shewhart Control Chart Constant (found in table H)
$\bar{R}$ is the average range

The control limits for the variability, as represented by the ranges, are given by

Sample #	X1	X2	X3	X4	X5	Σ X	Avg	R
1								
2								
3								
4								
5								
6								
7								
8								
9								
10								
11								
12								
13								
14								
15								
16								
17								
18								
19								
20								
21								
22								
23								
24								
25								
26								
27								
28								
29								
30								
31								
32								
33								
34								
35								
36								
37								
38								
39								
40								
41								
42								
43								
44								
45								
46								
47								
48								
49								
50								

Figure 7.4   Data Sheet for Collecting Variables Control Chart Measurements

$$UCL_R = D_4\overline{R}$$
$$LCL_R = D_3\overline{R}$$

where

$\overline{R}$ is the average range
D_4 is a Shewhart control chart constant
D_3 is a Shewhart control chart constant

If the subgroup size is 10 or greater it is usually recommended to use the sample standard deviation for the measure of variability. When this is the case the control limits for sample averages is as follows:

$$UCL_{\overline{x}} = \overline{\overline{x}} + A_3\overline{s}$$
$$LCL_{\overline{x}} = \overline{\overline{x}} - A_3\overline{s}$$

where

$\overline{s}$ is the average sample standard deviation
A_3 is a Shewhart constant

The limits for standard deviations are calculated using the following equations:

$$UCL_s = B_4\overline{s}$$
$$LCL_s = B_3\overline{s}$$

where

B_4 is a Shewhart Constant
B_3 is a Shewhart Constant

The limits for individuals are defined using the estimate for the standard deviation of the individuals. The limits are those between which virtually all, or 99.73 percent, of the individual values are expected to fall when only random variation is present. These limits are given by

$$UNTL = \overline{\overline{x}} + 3\overline{R}/d_2$$
$$LNTL = \overline{\overline{x}} - 3\overline{R}/d_2$$

where
> UNTL is the upper natural tolerance limit
> LNTL is the lower natural tolerance limit

9. Plot the control limits on graphs. These graphs should include space for identification of the characteristic being charted, identification of the sample and the time at which it was measured, and the notation of the actual control limits as calculated from the sample data. A sample graph is shown in Figure 7.5.

10. Issue the graphs to the inspectors or operators with instruction as to when the inspections should be made and which characteristics should be charted.

11. Instruct the inspectors or operators to notify quality or take other appropriate actions whenever a plotted point, whether an average, range, standard deviation, or individual, falls outside of the control limits shown on the graph or chart or whenever one of the "Western Electric" rules is noted. *The out-of-control points indicates that a special cause of variation is present and that the process is not the same as it has always been.*

12. Update the control limits periodically. All production processes exhibit some changes over time. Thus the process should be regularly monitored to determine when the control limits need to be changed. The use of new computer technology facilitates this periodic updating of the control limits, as well as assisting with the interpretation of the control limits and the existing process variability.

Example 7.3

The inspector at Aft-Tech's plate assembly operation has recorded the data shown in Figure 7.6 in order to establish control limits on a critical length dimension. The data that have been recorded reflect deviations from the desired dimension.

As can be seen from the data, the sample size specified was 5. Ten samples of 5 were taken and the critical length dimension was measured to complete the data collection for the base period, or precontrol analysis. Figure 7.6 also shows the sample average and the sample range for each sample. Develop control charts based on this information.

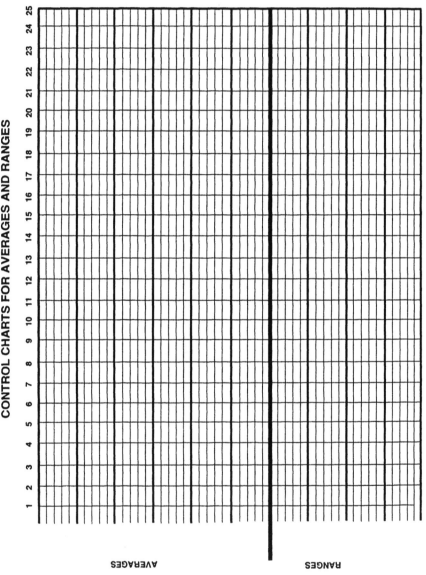

Figure 7.5   Sample Control Chart for Averages and Ranges

Sample #	X1	X2	X3	X4	X5	ΣX	Avg	R
1	1	0	0	2	-2	1	0.2	4
2	-1	0	-1	-1	-1	-4	-0.8	1
3	-1	2	0	-2	-1	-2	-0.4	4
4	-1	1	-1	0	0	-1	-0.2	2
5	-3	-3	2	-1	1	-4	-0.8	5
6	2	3	3	0	0	8	1.6	3
7	1	2	-1	-1	0	1	0.2	3
8	1	2	0	0	0	3	0.6	2
9	1	-1	0	1	-1	0	0	2
10	-2	0	2	-3	-2	-5	-1	5

Figure 7.6 Base Period Data

Solution

Once $\bar{x}$ and R have been calculated for each sample, the next step in the stated methodology is to calculate the process average, $\bar{\bar{x}}$, and the average range, $\bar{R}$. The sum of the 10 sample averages is –.6, and the average of these averages, the average average, is –.6/10 = –.06. The sum of the sample ranges is 31, and the average of the sample ranges is 31/10 = 3.1.

The next step in the control chart methodology is calculation of the control limits for averages and ranges. For averages the formulas and calculations are

$$UCL_{\bar{x}} = \bar{\bar{x}} + A_2\bar{R} = -.06 + (.58)(3.1) = 1.74$$
$$LCL_{\bar{x}} = \bar{\bar{x}} - A_2\bar{R} = -.06 - (.58)(3.1) = -1.86$$

The control limits for ranges are

$$UCL_R = D_4\bar{R} = (2.11)(3.1) = 6.54$$
$$LCL_R = D_3\bar{R} = (0)(3.1) = 0$$

If limits for individuals are desired the estimated standard deviation must be calculated. This is done with the equation

$$\sigma' = \bar{R}/d_2 = 3.1/2.326 = 1.33$$

The control limits for individuals are calculated as follows:

$$UNTL = \bar{\bar{x}} + 3\sigma' = -.06 + (3)(1.33) = +3.93$$

$$LNTL = \bar{\bar{x}} - 3\sigma' = -.06 - (3)(1.33) = -4.05$$

In order to be useful, these results need to be interpreted. If only random variation is present:

- Virtually all of the average values should fall between +1.74 and −1.86.
- Virtually all of the sample ranges should fall between 0 and 6.54.
- Virtually all of the individual values should fall between +3.93 and −4.05.

If any values were found outside the respective limits, it would be logical to assume that a special assignable cause of variation had entered the process.

Step 9 in the methodology is to plot the limits that have been calculated. These control limits show the expected limits of process variability. Figure 7.7 shows the control limits and averages (central lines) for each measured characteristic. The y axis is an actual measurement or quality attribute. The x axis identifies the sample that is charted, usually by the date/time that the sample is gathered.

The control limits should be identified very clearly so that the individuals posting the points will recognize when the process goes out of control. The statistics show when the process has changed — when assignable variation has entered the process. It then becomes important to determine the special cause of variation and either remove it or replicate it as necessary.

Example 7.4

The charts developed in Example 7.3 were issued to production. Table 7.1 shows the inspection data recorded for the operation for one day. The sheet shows the individual values, the sample averages, and the sample ranges for the eight hourly samples of five that were specified on the in-process inspection sheet. Plot and interpret these values.

Solution

The averages and ranges are plotted on the appropriate control charts in Figure 7.8. Normally, charts in this type of application are not kept for individuals.

An examination of the chart for averages, the $\bar{x}$ chart, shows that the third sample average is out of control. It should be concluded that assignable

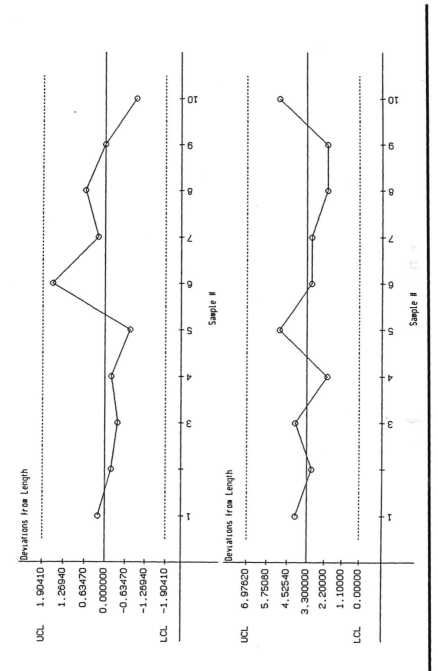

Figure 7.7 Control Chart for Example 7.3

Table 7.1

Sample	1	2	3	4	5	x	R
1	2	2	−1	2	−3	0.4	5
2	1	−3	1	−2	2	−0.2	5
3	3	3	1	1	1	1.8	2
4	−1	1	−1	2	0	0.2	3
5	−2	2	−1	0	5	0.8	7
6	−4	2	−2	1	0	−0.6	6
7	−1	2	−1	2	0	0.4	3
8	−3	2	1	2	3	1.0	6

variation is present. The range of the fifth sample is also out of control, again indicating assignable variation, as is an individual value in sample 6. Whenever a single point falls outside of the control limits, assignable variation is indicated and corrective action should be taken immediately.

Additional Out of Control Indications

Although a single point outside of the control limits immediately signals an out of control process, there are some other indications that are commonly used to identify assignable or abnormal variation. Duncan presents five additional criteria for identifying out of control processes (Duncan, 1974, p. 392). There are many other criteria that can be used as well. Sometimes these criteria are referred to as "Western Electric Rules". This is based on guidelines suggested in the original Western Electric Quality Control Handbook.

1. One or more points in the vicinity of the two sigma warning limits.
2. A run of seven or more points. (The run may be seven consecutive ascending points, seven consecutive decreasing points, or seven points above or below the mean.)
3. Cyclical patterns of points.
4. A run of three points outside the two sigma limits.
5. A run of five points outside the one sigma limits.

Note: One and two sigma control limits can be calculated using other constants provided by Dr. Shewhart.

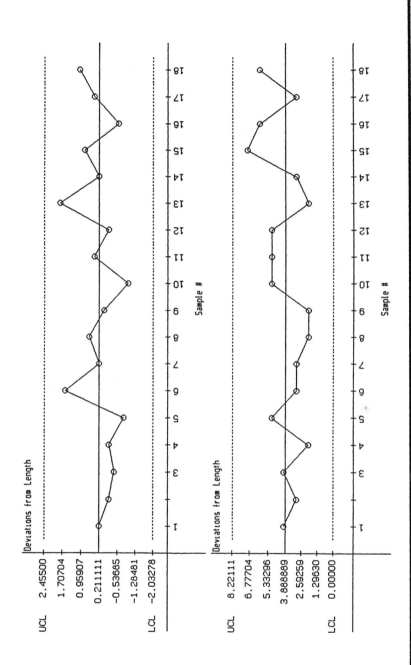

Figure 7.8 Control Chart for Example 7.4

The following examples will clarify the process of constructing and using control charts for variables. Much of the data will be presented on typical quality control inspection forms rather than within the written text of the problem. It must be pointed out that in actual practice, the data would normally be charted as they were collected, rather than after the fact.

Example 7.5

Aft-Tech's hot rod division has indicated a need to establish and maintain control charts for a cutoff operation on a certain rod that is processed by the division. A sample size of seven was selected for the base period analysis. Eight samples of seven provide sufficient data to establish the initial control limits. These 56 measures of rod length after the saw cut are recorded in Figure 7.9. The average and the range for each sample is calculated and entered on the data collection form. Develop control charts for averages, ranges, and individuals based on this information.

Sample Identification	SAMPLE NUMBER										SUM	AVERAGE X̄	RANGE R
	1	2	3	4	5	6	7	8	9	10			
1	5.2	4.9	5.1	5.8	4.7	4.2	4.4				34.3	4.9	1.6
2	5.3	4.7	2.8	3.4	4.2	3.9	4.1				28.4	4.1	2.5
3	3.1	4.4	2.6	2.9	4.0	2.7	2.8				22.5	3.2	1.8
4	3.0	3.1	3.1	2.9	4.3	4.0	3.2				23.6	3.4	1.4
5	3.1	5.0	2.6	2.8	2.8	3.1	3.3				22.7	3.2	2.4
6	4.6	4.4	2.8	3.0	5.1	3.1	2.9				25.9	3.7	2.2
7	5.0	4.9	2.8	2.9	4.3	3.1	2.9				25.9	3.7	2.2
8	2.9	5.3	3.1	3.0	3.1	4.6	3.2				25.2	3.6	2.4
											Sum	29.8	16.5

Figure 7.9   Base-Period Data

Solution

The value for σ' is determined.

$$\sigma = \overline{R}/d_2 = 2.1/2.704$$

The limits are computed using the appropriate Shewhart control chart constants from Table H.

Averages:

$$UCL_{\overline{x}} = 3.73 + (.42)(2.1) = 4.61$$
$$LCL_{\overline{x}} = 3.73 - (.42)(2.1) = 2.85$$

Ranges:

$$UCL_R = (1.92)((2.1) = 4.03$$
$$LCL_R = (.08)(2.1) = 0.17$$

Individuals:

$$UNTL = 3.73 + (3)(.78) = 6.07$$
$$LNTL = 3.73 - (3)(.78) = 1.39$$

Figure 7.10 shows the two sets of limits that have been calculated in this base period analysis for the respective measurement — averages and ranges. These are the limits between which it is expected that all of the values will normally fall when only random variation is present.

Example 7.6

After the charts in example 7.5 had been prepared, they were put into use. Inspection data were collected in the same manner as in the base period analysis. Figure 7.11 shows the in-process inspection instructions. Figure 7.12 shows inspection data for the hod rod division for the first day the control chart was in use. The sample averages and the sample ranges have been calculated for each of the 16 inspections. Plot and interpret these values.

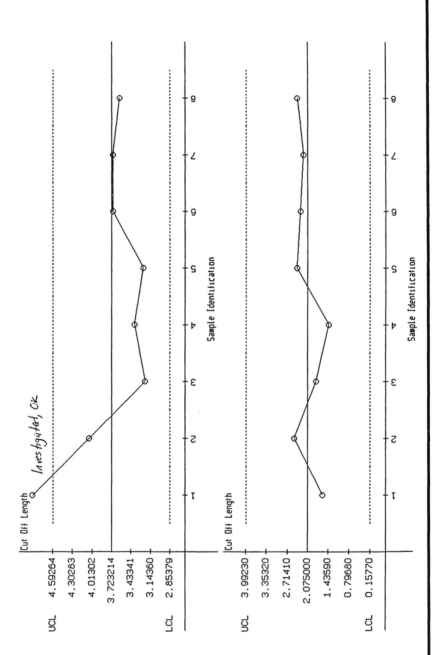

Figure 7.10　Control Chart for Example 7.5

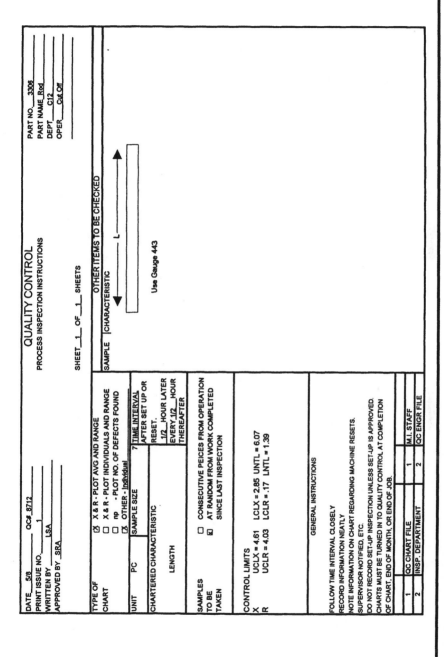

Figure 7.11 In-Process Inspection Sheet

CONTROL CHART DATA COLLECTION FORM	Page _/_ of _/_

Date 9/20   Operation No. cut-off

Sample Identification	SAMPLE NUMBER										SUM	AVERAGE X̄	RANGE R
	1	2	3	4	5	6	7	8	9	10			
8:00 AM	4.1	4.1	4.3	4.2	4.2	4.5	4.2				29.6	4.2	0.4
8:30 AM	4.3	3.6	2.9	4.9	3.9	4.2	4.0				27.8	4.0	2
9:00 AM	2.8	5.2	4.3	3.7	4.7	5.0	3.0				28.7	4.1	2.4
9:30 AM	5.2	4.8	3.5	4.2	5.0	4.5	1.9				29.1	4.2	3.3
10:00 AM	4.2	5.2	3.8	3.9	4.7	3.9	2.0				27.7	4.0	3.2
10:30 AM	3.2	4.1	2.9	4.5	3.6	3.7	4.1				26.1	3.7	1.6
11:00 AM	4.5	4.2	4.2	4.1	5.2	3.3	3.4				28.9	4.1	1.9
11:30 AM	4.0	3.9	3.9	4.3	4.3	4.8	4.2				29.4	4.2	0.9
12:00 PM	3.0	4.2	4.7	2.8	3.7	2.9	3.1				24.4	3.5	1.9
12:30 PM	4.1	5.0	5.0	5.1	4.7	2.6	2.5				29.0	4.1	2.6
1:00 PM	4.3	4.5	4.7	4.2	5.0	3.1	3.0				28.8	4.1	2
1:30 PM	3.7	3.9	3.7	4.5	3.0	3.6	3.4				25.8	3.7	1.5
2:00 PM	3.8	3.7	2.7	4.0	2.8	2.8	3.6				23.4	3.3	1.3
2:30 PM	3.6	4.5	3.8	3.0	4.0	3.7	3.3				25.9	3.7	1.5
3:00 PM	3.9	4.0	2.0	4.3	4.6	4.5	4.2				27.5	3.9	2.6
												58.8	

Figure 7.12   Inspection Data

Solution

The points are plotted on the control chart in Figure 7.13. Analysis of the charts shows that the averages, ranges, and individuals are all in control. Since there are not special causes of variation present, it can be said that the process is in control. Additional data must continually be collected since special causes of variation may arise at any time.

Example 7.7

The Aft-Tech pop top manufacturing division decided to establish control limits for the diameters of the holes in the pop top rings. A sample size of four was selected, and 13 samples or subgroups were inspected, for a total of 52 readings. These data, along with the sample averages and ranges, are shown in Figure 7.14. Develop control charts for averages and ranges based on these data.

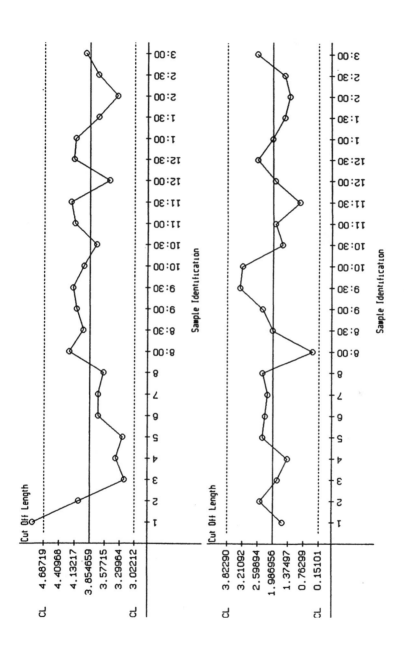

Figure 7.13 Control Chart for Example 7.6

Sample #	X_1	X_2	X_3	X_4	ΣX	Avg	Range
1	0.22	0.25	0.19	0.24	0.90	0.23	0.06
2	0.21	0.27	0.20	0.23	0.91	0.23	0.07
3	0.24	0.23	0.25	0.23	0.95	0.24	0.02
4	0.23	0.24	0.24	0.21	0.92	0.23	0.03
5	0.26	0.19	0.22	0.21	0.88	0.22	0.07
6	0.24	0.23	0.23	0.24	0.94	0.24	0.01
7	0.23	0.25	0.26	0.21	0.95	0.24	0.05
8	0.24	0.24	0.26	0.21	0.95	0.24	0.05
9	0.20	0.26	0.24	0.23	0.93	0.23	0.06
10	0.22	0.25	0.26	0.24	0.97	0.24	0.03
11	0.23	0.25	0.24	0.22	0.94	0.24	0.03
12	0.23	0.24	0.25	0.25	0.97	0.24	0.20
13	0.24	0.21	0.21	0.22	0.88	0.22	0.03

Totals   3.03   0.53

Figure 7.14   Base-Period Data

Solution

The average average, $\bar{\bar{x}}$, is calculated from these data:

$$\bar{\bar{x}} = 3.03/13 = .233$$

The average range, $\bar{R}$, is calculated:

$$\bar{R} = .53/13 = .041$$

The control limits are calculated:

Averages:

$$\text{UCL}_{\bar{x}} = .233 + (.73)(.041) = .263$$
$$\text{LCL}_{\bar{x}} = .233 - (.73)(.041) = .203$$

Ranges:

$$\text{UCL}_R = (2.26)(.041) = .093$$
$$\text{LCL}_R = (0)(.041) = 0$$

The control charts are shown in Figure 7.15.

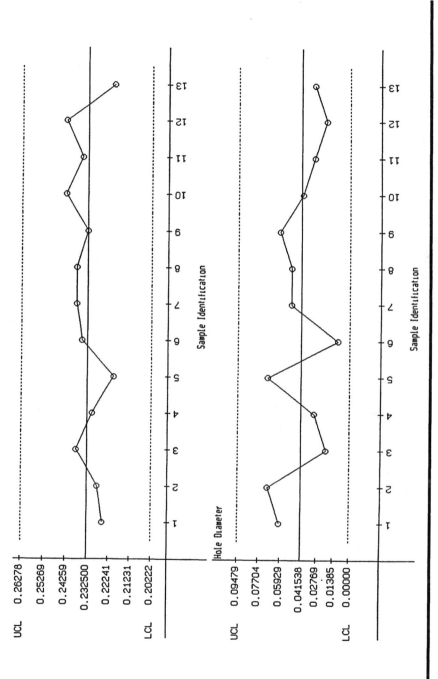

Figure 7.15 Control Chart for Example 7.7

Example 7.8

The charts developed in Example 7.7 were put into use to monitor the process. A sampling frequency of one time per day was judged to be reasonable for keeping tabs on this process. Two weeks' worth of data are shown in Figure 7.16. Plot and interpret these values.

Sample #	X1	X2	X3	X4	ΣX	Avg	Range
9-1	0.22	0.23	0.19	0.25	0.89	0.22	0.06
9-2	0.24	0.25	0.22	0.21	0.92	0.23	0.04
9-3	0.23	0.25	0.26	0.26	1.00	0.25	0.03
9-4	0.25	0.23	0.24	0.25	0.97	0.24	0.02
9-5	0.24	0.26	0.25	0.24	0.99	0.25	0.02
9-6							
9-7							
9-8	0.23	0.25	0.26	0.22	0.96	0.24	0.04
9-9	0.26	0.21	0.19	0.28	0.95	0.24	0.10
9-10	0.21	0.24	0.23	0.24	0.92	0.23	0.03
9-11	0.24	0.25	0.25	0.23	0.97	0.24	0.07
9-12	0.23	0.28	0.21	0.22	0.94	0.24	0.07

Figure 7.16 Additional Inspection Measurements

Solution

The sample averages and ranges are plotted on the control charts shown in Figure 7.17.

The results from the sample taken on September 9 show that the range is out of control, although on that date the averages are in control. This out-of-control point indicates that there is some assignable variation present in the process. Something is happening to cause the process to vary more or differently than would normally be expected. Investigation is required to determine the special cause or causes. Once the cause is identified and steps taken to prevent it from recurring, the process will return to a state of statistical control as long as no additional special causes of variation enter the process.

Significance of the Control Chart Constants

At this point a brief digression into one of the statistical interpretations of control limits for averages is appropriate. When control limits for averages

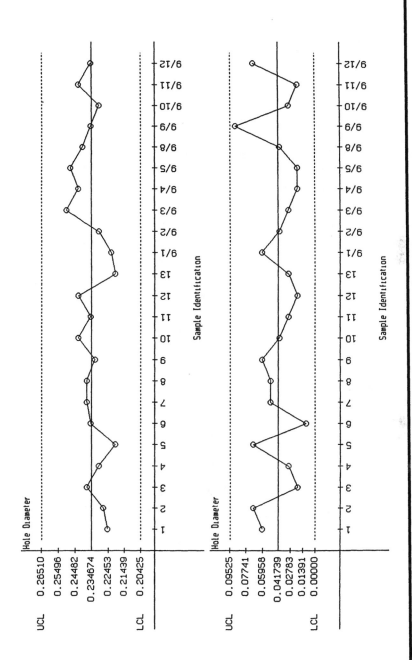

Figure 7.17 Control Chart for Example 7.8

are being calculated there may be some question about the significance of the Shewhart control chart constants. To illustrate exactly what the value of A_2 means, it is necessary to point out that the upper control limit for averages, which is calculated by multiplying A_2 times $\overline{R}$, is the three sigma limit. That means that $(A_2)(\overline{R}) = 3$ standard deviations for the distribution of sample averages.

Control Charts for Individuals and Moving Ranges

Some variables do not vary significantly during a short period of time. If we were to take a measurement of the characteristic now, and then again two minutes from now, we would not expect to see any difference. For example, if we were measuring the humidity in a room, we would expect to see no variation between now and 2 minutes from now. Instead, we would more realistically expect to find variation over a longer period of time. The humidity might change from the present time and a time two hours, rather than two minutes, from now. This indicates that variation from sample to sample is more representative than the variation within a given sample.

For some characteristics, in order to measure the characteristic the product must be destroyed. This could be quite costly and might, in the extreme case, cause us to have minimal product to ship after in-process checks.

If this is the case, of if between-sample variation is more important than within-sample variation, then we must develop control charts that will show this: the individual and moving range chart.

In an individual and moving range chart, the individual chart shows the measurement values associated with each sample value and the range chart shows the differences between successive samples.

The control limits for individuals usually show the values between which we expect virtually all of the individual values to fall. The control limits for ranges show the limits within which we normally expect to find the variability between samples to fall.

The formulas for these limits are as follows:

Individuals

$$UCL_x = \overline{x} + (3)(\overline{R})/d_2$$
$$LCL_x = \overline{x} - (3)(\overline{R})/d_2$$

Moving ranges

$$UCL_{MR} = (D_4)(\bar{R})$$
$$LCL_{MR} = (D_3)(\bar{R})$$

For each set of equations we use n = 2 to access the Shewhart constants in Table H.

Example 7.9

A bakery is monitoring the temperatures that the ovens baking the name brand cookies are maintaining. The recipe calls for a temperature of 350 degrees. It is known that cookies are acceptable to customers if the temperature varies by no more than 3 degrees in either direction.

Over a period of two weeks, the bakery collected the data shown in Figure 7.18. Determine control limits for the individuals and moving ranges.

Solution

The process average, x, is calculated to be 350.5625. The between-sample range is shown in Figure 7.19. The range is the difference between successive sample measurements. The average of these values for the sample range is 1.4839.

This is referred to as a two-sample moving range. It is possible to look for the variation between every third sample, in which case we would have a three-sample moving range. It is also possible to look for the variation between every fourth sample. It is obviously more difficult to determine this difference the more samples that are skipped over. We normally stick with the simplest case of the two-sample moving range.

Using the sample size of n = 2, the limits are calculated using the equations defined above.

$$UCL_x = 354.51$$
$$LCL_x = 346.61$$
$$UCL_{MR} = 4.85$$
$$LCL_{MR} = 0$$

Date	Time	Temperature
12-1	8:00	350
12-1	10:00	352
12-1	12:00	349
12-1	2:00	351
12-2	8:00	351
12-2	10:00	352
12-2	12:00	350
12-2	2:00	351
12-3	8:00	352
12-3	10:00	351
12-3	12:00	351
12-3	2:00	351
12-4	8:00	349
12-4	10:00	351
12-4	12:00	350
12-4	2:00	351
12-5	8:00	350
12-5	10:00	350
12-5	12:00	349
12-5	2:00	352
12-8	8:00	351
12-8	10:00	352
12-8	12:00	348
12-8	2:00	352
12-9	8:00	351
12-9	10:00	351
12-9	12:00	350
12-9	2:00	353
12-10	8:00	349
12-10	10:00	352
12-10	12:00	348
12-10	2:00	352
12-11	8:00	351
12-11	10:00	351
12-11	12:00	349
12-11	2:00	352
12-12	8:00	351
12-12	10:00	350
12-12	12:00	348
12-12	2:00	350

Figure 7.18   Bakery Time and Temperature Data

Date	Time	Temperature	Range
12-1	8:00	350	---
12-1	10:00	352	2
12-1	12:00	349	3
12-1	2:00	351	2
12-2	8:00	351	0
12-2	10:00	352	1
12-2	12:00	350	2
12-2	2:00	351	1
12-3	8:00	352	1
12-3	10:00	351	1
12-3	12:00	351	0
12-3	2:00	351	0
12-4	8:00	349	2
12-4	10:00	351	2
12-4	12:00	350	1
12-4	2:00	351	1
12-5	8:00	350	1
12-5	10:00	350	0
12-5	12:00	349	1
12-5	2:00	352	3
12-8	8:00	351	1
12-8	10:00	352	1
12-8	12:00	348	4
12-8	2:00	352	4
12-9	8:00	351	1
12-9	10:00	351	0
12-9	12:00	350	1
12-9	2:00	353	3
12-10	8:00	349	4
12-10	10:00	352	3
12-10	12:00	348	4
12-10	2:00	352	4
12-11	8:00	351	1
12-11	10:00	351	0
12-11	12:00	349	2
12-11	2:00	352	3
12-12	8:00	351	1
12-12	10:00	350	0
12-12	12:00	348	2
12-12	2:00	350	2

Figure 7.19 Moving Range Calculations for Bakery Data

This means that we would expect 99.73 percent of the individual values to fall between 354.51 and 346.61. We also would expect virtually all of the between-sample variation to fall between 0 and 4.85. As long as additional samples have individuals between these limits and differences between these limits, we are confident that the process is performing as consistently as it always has.

Figure 7.20 shows these limits drawn on the appropriate control charts. When using this type of chart it is essential to remember that we are concerned with *between-sample* variation rather than *within-sample* variation.

Control Charts for Averages and Standard Deviations

Sometimes, especially when the sample size becomes as large as 10, using the range as a measure of variability is subject to distortion from outliers. When this is the case we need to use the sample standard deviation as the measure of variability. The procedure we use to establish the control charts is almost the same as the one we used for establishing the $\bar{x}$ and R charts, *with the exception that instead of calculating the range for each sample we calculate the sample standard deviation for each sample and we use the average standard deviation, s, instead of the average range.* The control limit formulas are calculated as follows:

Averages

$$UCL_{\bar{x}} = x + (A_3)(\bar{s})$$
$$LCL_{\bar{x}} = x - (A_3)(\bar{s})$$

Standard Deviations

$$UCL_s = (B_4)(\bar{s})$$
$$LCL_s = (B_3)(\bar{s})$$

The interpretation of the chart is the same whether ranges or standard deviations are used to measure the within sample variation.

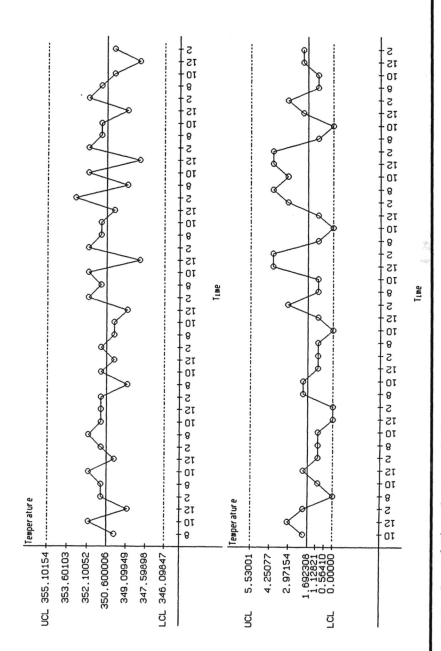

Figure 7.20   Control Chart for Example 7.9

Example 7.10

Figure 7.21 shows some sample data. Calculate $\bar{x}$ and s control limits.

Sample	Value 1	Value 2	Value 3	Average	Standard Deviation
1	8	6	6	6.67	1.15
2	7	6	9	7.33	1.53
3	8	8	4	6.67	2.31
4	5	7	7	6.33	1.15
5	5	9	4	6.00	2.65
6	8	9	10	9.00	1.00
7	7	9	5	7.00	2.00
8	7	8	8	7.67	0.58
9	9	10	10	9.67	0.58
10	11	7	6	<u>8.00</u>	<u>2.65</u>
				7.43	1.56

Figure 7.21

Solution

Using the values calculated above and the Shewhart constants from Table H the control limits are calculated as follows:

$$UCL_{\bar{x}} = 7.43 + (1.954)(1.56) = 10.478$$

$$LCL_{\bar{x}} = 7.43 - (1.954)(1.56) = 4.382$$

$$UCL_{s} = (2.568)(1.56) = 4.006$$

$$LCL_{s} = 0$$

Control Charts for Attributes

The control charts examined up to this point have been control charts for variables, or characteristics that could be measured. When the use of linear measurements in not practical, then attributes are often counted and appropriate charts constructed. Some of the more common attribute control charts are number of defects per sample (c charts), number of defects per unit (u charts), and percentage defective charts (p charts.)

A **defect** can be defined as "a departure of a quality characteristic from its intended level or state that occurs with a severity sufficient to cause an associated product or service not to satisfy intended normal, or reasonably foreseeable, usage requirements" (ASQC, 1983, p.13).

A **defective,** on the other hand, is "a unit of product or service containing at least one defect, or having several imperfections that in combination cause the unit not to satisfy intended normal, or reasonably foreseeable, usage requirements" (ASQC, 1983, p. 15).

u Charts

"The control chart for u is most useful when several independent nonconformities (they must be independent) may occur in one unit of a product" (Juran, 1988, p. 24.22). This is most likely to occur in complex assemblies.

Control limits for the u chart are calculated as follows: The mean, $\bar{u}$, is calculated as the total number of defects divided by the total number of units checked. The upper and lower limits depend on each specific sample size. As such, the limits may vary from sample to sample.

$$UCL_u = \bar{u} + 3\sqrt{\bar{u}/n}$$

$$LCL_u = \bar{u} - 3\sqrt{\bar{u}/n}$$

Example 7.11

Aft-Tech's Customer Service Department wanted to determine the control limits for the number of errors service writers were making in completing service forms. The following data were collected:

Service Writer	No. Forms	No. Errors	Errors/Form
Walter	28	50	1.79
Steve	17	43	2.53
Dana	44	60	1.36
John	30	45	1.50
Mary	30	40	1.33
Peter	30	48	1.60
Sam	30	44	1.47
Larry	30	15	0.50
Susan	44	44	1.00
Sum	283	389	

Solution

The average number of errors per form is calculated.

$$\bar{u} = 389/283 = 1.37$$

The control limits must be calculated for each sample that has a different sample size. The values of $\sqrt{\bar{u}/n}$ and the control limits for each point are shown below.

Sample	No. Forms	$(3)\left(\sqrt{\bar{u}/n}\right)$	UCL	LCL
Walter	28	.66	2.03	0.71
Steve	17	.85	2.22	0.52
Dana	44	.53	1.90	0.84
John	30	.64	2.01	0.73
Mary	30	.64	2.01	0.73
Peter	30	.64	2.01	0.73
Sam	30	.64	2.01	0.73
Larry	30	.64	2.01	0.73
Susan	44	.53	1.90	0.84

The control chart for these points is shown in Figure 7.22.

c Charts

The c chart is the control chart for the number of defects. This chart is primarily used when the samples are the same size. "It is particularly effective when the number of nonconformities possible on a unit is large but the percentage for any single nonconformity is small." (Juran, 1988, p. 24.23) The control limits for c charts depend on the average number of defects per sample, c. They are calculated as follows:

$$UCL_c = \bar{c} + 3\sqrt{\bar{c}}$$

$$LCL_c = \bar{c} - 3\sqrt{\bar{c}}$$

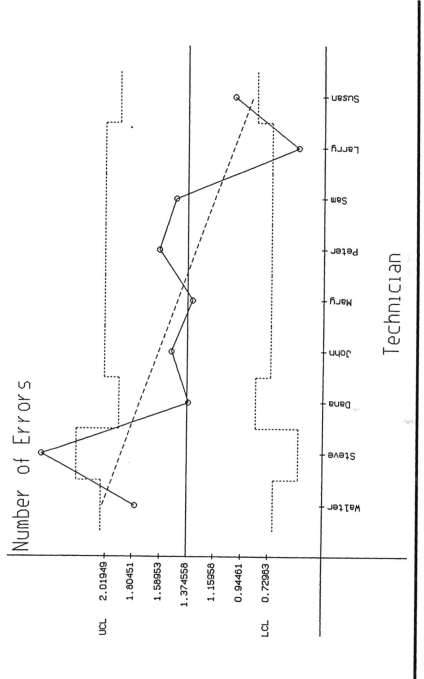

Figure 7.22 Control Chart for Example 7.11

Example 7.12

The Aft-Tech weaving division has just received the information shown in Table 7.2 from its inspection reports. Aft-Tech wants to establish control limits for this process. Since each lot represents a significant amount of production, the data base is large enough for establishment of control limits.

Table 7.2

Lot	Number of Defects
1	60
2	80
3	70
4	60
5	90
6	85
7	77
8	80
9	65
10	72
11	70
12	65
	874

Solution

The values for the number of defects per lot are known, so the average of these, $\bar{c}$, can be calculated.

$$\bar{c} = 874/12 = 72.83 \text{ defects/lot}$$

The control limits are calculated from the formulas presented above.

$$\text{UCL}_c = 72.83 + (3)\sqrt{(72.83)} = 98.43$$

$$\text{LCL}_c = 72.83 - (3)\sqrt{(72.83)} = 47.23$$

If only random variation is present, Aft-Tech can expect to find between 47.23 and 98.43 defects per lot produced in its weaving division. Any more or fewer defects indicates the presence of special causes of variation.

If the percent of possible values desired was something other than the 99.73 percent, or "virtually all" included in the three sigma limits, a coefficient other than three could be used. For example, two sigma limits would include about 95.5 percent of the potential values.

p Charts

The percentage defective or proportion defective classification is used in cases where an item, after inspection, is classified as either acceptable or unacceptable. The percentage defective, p, is the number of defective units in a sample of size n, divided by the sample size. The average percentage defective, p, is the sum of the total number of defectives divided by the total sample size. Control limits are calculated as follows:

$$UCL_p = \bar{p} + (3)\sqrt{(\bar{p})(1-\bar{p})/n}$$
$$LCL_p = \bar{p} - (3)\sqrt{(\bar{p})(1-\bar{p})/n}$$

As with the u chart, when the sample size varies the control limits will vary. The larger the sample size the closer to the mean the control limits will be.

Example 7.13

The Aft-Tech minibike quality control organization has tallied the information in Table 7.3 from last week's inspection reports and wishes to establish p charts for the bikes.

Table 7.3

Day	Number Produced	Number Rejected	Percentage
Monday	150	14	0.093
Tuesday	150	19	0.126
Wednesday	150	16	0.106
Thursday	150	18	0.120
Friday	150	15	0.100
	750	82	

Solution

The daily percentage defective is included in the inspection results. The average percentage defective is calculated

$$\bar{p} = 82/750 = .109$$

Because all of the samples are the same size the limits will all be the same. They are calculated as follows:

$$UCL_p = .109 + (3)\sqrt{(.109)(.891)/150} = .1852$$

$$LCL_p = .109 - (3)\sqrt{(.109)(.891)/150} = .0328$$

Virtually all of the daily percentages defective, based on samples of size 150, would be expected to fall within these limits. The presence of a daily percentage outside of these limits (as long as the sample size remained 150 — if it did not, a new set of limits would have to be calculated) would indicate the presence of a special cause of variation.

np Charts

The np chart, also known as the number of nonconforming, is closely related to the p chart. The additional constraint of using an np chart is that the sample size *must* be uniform; np is the direct count or actual number of nonconforming units in each subgroup or sample.

Calculation of the control limits is as follows: The average number of nonconforming, np, is calculated. This is the total number of nonconforming units divided by the total in all of the subgroups. The control limits are calculated using these formulas as long as the average percentage defective is approximately .1 or less:

$$UCL_{np} = n\bar{p} + 3\sqrt{n\bar{p}}$$

$$LCL_{np} = n\bar{p} - 3\sqrt{n\bar{p}}$$

Example 7.14

Use the data in Table 7.3 to calculate np limits.

Solution

The average percentage defective for this material is .109, which is close enough to .1 to use the above-shown control limit formula. $n\bar{p}$ is calculated to be 82/5 = 16.4 The control limits are calculated:

$$UCL_{np} = 16.4 + (3)\sqrt{(16.4)} = 28.55$$

$$LCL_{np} = 16.4 - (3)\sqrt{(16.4)} = 4.25$$

Cumulative Sum Control Charts

A third type of control chart, in addition to the charts for variables and the charts for attributes, is the cumulative sum control chart, or the cusum chart. Developed by E. S. Page in 1954, it has been employed particularly in England (Johnson, 1962, p. 15).

"The cumulative sum control chart... is used primarily to maintain current control of a process. Its advantage over the ordinary Shewhart chart is that it may be equally effective and less expense." (Duncan, 1974, p. 464) Since illustration of the use of cusum charts is beyond the scope of this book, the cusum chart will merely be compared with the Shewhart control chart, which has been explained and illustrated in some detail throughout this chapter.

The cusum chart was designed to show changes in a process characteristic, such as mean or percentage defective, whereas the Shewhart control chart is more concerned with the consistency of a process. On Shewhart charts, each point describes a specific sample. Each of these samples is individually compared with the control limits to determine the state of control. Although some consideration may be given to past sample values, such as through the use of runs, the Shewhart chart analysis is based primarily on the results of each sample inspection. In the cusum chart, "Starting from a given point, all subsequent plots contain information from the whole of observations, up to and including the plotted point" (Johnson, 1962, p. 17). This cumulative result is compared with the cusum chart's version of a control limit in order to determine whether or not the charted characteristic has displayed a significant change.

Many of the applications of cusum charts have been in the chemical industries (Truax, 1961, pp. 18–25). The cumulative sum control chart will not replace the Shewhart control chart, but it does have suitable applications and thus should be considered as a potential analysis procedure.

Summary

Control charts tell when a process is out of control or not producing as it has in the past. Comparison of control chart data with specifications will reveal whether or not the process is producing acceptable material. Chapter 8 will discuss the concept of process capability in more detail.

When a process is out of control, special causes of variation have entered the process. Problem solving skills must be used to determine the assignable cause of this variation. Figure 7.23 shows the action to take on the basis of control chart data.

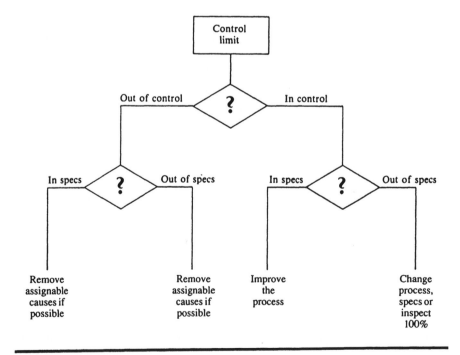

Figure 7.23   Action to be Taken in Conjunction with Control Charts

References

ASQC, *Glossary and Tables for Statistical Quality Control,* Quality Press, Milwaukee, 1983.

Duncan, A.J., *Quality Control and Industrial Statistics,* Irwin, Homewood, IL, 1974.

Johnson, N.L. and Leone, F.C., "Cumulative Sum Control Charts: Mathematical Principles Applied to Their Construction," *Industrial Quality Control,* June, 1962.

Juran, J. (Ed.), *Juran's Quality Control Handbook,* McGraw-Hill, New York, 1988.

Truax, H. Mack, "Cumulative Sum Charts and Their Application to the Chemical Industry," *Industrial Quality Control,* December, 1961.

Practice Problems

1. Calculate control limits for the data in Table 7.4.

Table 7.4

Sample	X_1	X_2	X_3	X_4	X_5	X_6	X_7	X_8
1	2	3	3	5	.3	1	2	6
2	3	2	5	3	4	3	2	4
3	1	2	3	4	3	2	5	2
4	3	4	5	6	3	3	3	3
5	2	3	4	7	2	2	1	4
6	1	1	2	2	3	3	4	4
7	3	2	2	7	3	2	3	4

2. The average of 1200 samples of size 9 was determined to be 125. The average range was 17.82. The sample standard deviation was 6. The following sample of 9 was measured. Is the sample in control?

    120 132 124 130 118 132 135 121 127

3. The tensile strength of a certain steel is 17.8. The value was determined by finding the average of samples of size 6. The average range was 2.4.

    (a) Calculate control limits for averages and ranges.

    (b) Additional sample ranges of 1.4, 1.3, 1.5, and 1.2 were measured. Is the range in control?

    (c) The means of the samples in part (b) were 18.0, 19.2, 18.5, and 16.0. Are the averages in control?

4. Given the following information, calculate the control limits for averages, individuals, and ranges:

$$\bar{\bar{x}} = 72 \quad n = 4 \quad \bar{R} = 6$$

5. Given the following information, calculate the control limits:

$$\bar{\bar{x}} = 46.5 \quad n = 3 \quad UCL_R = 7.725$$

6. A process has an average of 86 and a variance of 35.111. If the sample size is 3, what are the control limits for averages and ranges?

7. A sample of 50 groups of size 7 had an average of 75. The average range of these 50 samples was 12. Determine the control limits for averages, individuals, and ranges.

8. One hundred samples of 15 had a total of 5600 defects. Compute three-sigma control limits.

9. An electrical parts manufacturer required 100 percent inspection. During a specified analysis period, 50 lots of 300 parts yielded a total of 600 defects. What are the control limits for the number of defects per lot?

10. A shirt manufacturer wanted to check to see if the number of buttonholes per shirt was normally distributed. She inspected 500 samples of 12 shirts and determined that there were 30,127 buttonholes. A shirt is supposed to have 6 buttonholes.
    (a) Were the inspection results to be normally expected?
    (b) How many shirts in a sample of 144 would be expected to have the wrong number of buttonholes?

11. An airplane can carry a maximum of 100 people. It has averaged 81 passengers per trip. Would it be unexpected to expect a flight to be full?

12. A city bus has averaged 42 passengers per run on a given route. What is the largest capacity bus that would ever normally be needed?

13. Determine the appropriate control limits for the data in Table 7.5.

14. Determine the appropriate control limits for the data in Table 7.6.

15. Determine the appropriate control limits for the data in Table 7.7

16. Determine the $\bar{x}$ and s chart limits for the data shown in Table 7.8.

Table 7.5

Day	Number of Repairs
1	14
2	13
3	16
4	15
5	16
6	12
7	14
8	15
9	15
10	16
11	13
12	14

Table 7.6

X_1	X_2	X_3	X_4
0.55	0.6	0.57	0.55
0.58	0.5	0.51	0.55
0.59	0.55	0.60	0.58
0.51	0.53	0.55	0.58
0.55	0.5	0.55	0.51
0.56	0.59	0.51	0.51
0.54	0.57	0.50	0.60
0.55	0.53	0.54	0.54
0.50	0.51	0.55	0.50
0.54	0.53	0.58	0.55
0.58	0.58	0.53	0.52
0.53	0.54	0.54	0.57
0.60	0.50	0.53	0.57

Table 7.7

Sample	Size	Number Defective
1	180	11
2	180	19
3	180	15
4	180	11
5	180	17
6	180	19
7	180	16
8	180	10
9	180	16
10	180	17
11	180	19
12	180	14
13	180	17
14	180	16

Table 7.8

Sample	X_1	X_2	X_3	X_4
1	7	5	5	7
2	6	6	7	5
3	9	7	6	6
4	3	3	8	8
5	6	4	9	7
6	9	8	7	9
7	6	7	4	4

17. Determine control limits for individuals and moving ranges for the following data:

Time	Pressure
8:00	1344
10:00	1356
12:00	1363
2:00	1352
8:00	1355
10:00	1347
12:00	1347
2:00	1335
8:00	1365
10:00	1354
12:00	1348
2:00	1356
8:00	1359
10:00	1363
12:00	1354
2:00	1335
8:00	1368
10:00	1351
12:00	1363
2:00	1355
8:00	1346
10:00	1337
12:00	1376
2:00	1358
8:00	1363
10:00	1364
12:00	1363
2:00	1351
8:00	1363
10:00	1346
12:00	1364
2:00	1368
8:00	1349
10:00	1361
12:00	1342
2:00	1343
8:00	1336
10:00	1345
12:00	1364
2:00	1349

18. The following is attribute data showing the number of defects found in each sample of material inspected. Calculate control limits for the following data:

Sample	Sample Size	Number of Defects
1	200	400
2	200	430
3	200	158
4	350	500
5	350	400
6	350	450
7	350	480
8	275	350
9	275	380

19. Calculate np control limits for the following data:

Week	Units	Defects
1	200	15
2	200	14
3	200	16
4	200	12
5	200	13
6	200	16
7	200	16
8	200	15
9	200	11

8 Process Capability

Introduction

All production processes are the combination of three major components — materials, machines, and people. All of these are subject to variation, both random and assignable.

Consider a typical machining process. The product produced by a machining process depends on the quality of the raw material that is machined. Different materials have different characteristics when machined; some are much easier to process than others. And even a single material will be subject to some variability in hardness, porosity, grain structure, and so on. There is little that can be done to control this random variation within certain acceptable limits. A second source of variation is the machine used in the process. No matter how precise the equipment is, there will be some natural or normal variation due to how the equipment functions with a particular piece of material or to some other uncontrollable reason. The third and the most variable element in the process is the person. People, by nature, are inconsistent. It is virtually impossible for a person to perform the same operation twice in exactly the same way.

Before the ability of a process to meet the desired standards can be defined, the process capability must be defined.

The **process capability** is the limits between which individual values produced by any process would normally be expected to fall when only random variation is present.

Traditional definitions of process capability have expressed it as 6σ, identifying where virtually all of the individual values produced by the process will fall. Once the process capability has been established, predictions can be made regarding the quality level of the material being produced.

Specifications

In Chapter 7 the concept of statistical control was developed. A process that is in statistical control is producing material that is not significantly different from material that has been produced in the past by that process. Although it is important to know whether a process is producing consistently, it is just as important to know whether the process is producing material as it should be. Does the material being produced meet the customers' requirements? Being in control means being consistent, but being consistently wrong is not a desirable condition.

Ultimately the quality control process must answer two questions. Is the process consistently producing material that performs as intended? Does the process create a product that meets the desired specifications?

> **Specification limits** are those limits placed on a product, product characteristic, or service, by engineers or designers for the purpose of ensuring the proper functioning of the finished product and for meeting customer requirements.

Specification, or specs, reflect the limits between which all individual measurements of characteristics or parameters *should* fall. The measurements of the characteristics must fall with the specs for the product to meet customer requirements. Unfortunately, not all processes produce material that meets or exceeds specifications.

When material produced does not meet specifications, the organization has three options. First, the nonconforming (defective) material may be separated from the rest. Second, a different, better, or more precise production process may be used, if one is available. Third, the specifications may be changed. Specs are not always realistically set by the designers. Sometimes, rather than reflecting the product requirements, the specifications merely reflect a traditional method of stating requirements. For example, some organizations specify that all linear dimensions must be maintained to within ± .005 cm, regardless of the use of the product.

The second and third options, better equipment and better specifications, are the preferable solutions, However, many organizations are obliged to follow the first path.

If the first option must be chosen, wouldn't it be nice to know ahead of time how many or what percentage of the parts or products produced will not conform to the quality specifications? If the percentage of defective material can be estimated, appropriate sampling plans can be designed to effectively screen the defective material from the acceptable material as long as the process is in a state of statistical control.

Determining Process Capability

There is a tool, or procedure, that uses basic statistical analysis and control chart theory to make the prediction of the expected percentage defective a fairly straight forward task. This procedure is often call the **process capability study** because it is used to analyze the ability of a process to produce material within certain limits. These are usually three-sigma limits for individuals, which we have called the natural tolerance limits. These limits are typically used since they encompass approximately 99.73 percent of all of the individual values. Some organizations have chosen to use other limits. A more common variation is the use of plus or minus six-sigma limits.

Once the control limits for individuals have been calculated, the process capability is known. If the variation permitted in the product specifications is greater than that in the natural tolerance limits, virtually no defective, substandard, or parts not meeting customer requirements will be produced. If on the other hand, the variation in process capabilities is greater than in the product specifications, a measurable or predictable amount of defective material will be produced.

Generally, organizations are satisfied when the variation in product specifications is greater than that in process capability. As long as virtually no defective material is being produced and the process is in control, there is no need to worry. But when the variation in process capability exceeds that in product specifications, organizations need to know how much material will not conform to specifications. This has an obvious and direct bearing on the cost of the material produced. The question can be answered by the statistical analysis procedure of calculating probabilities under the normal curve. The statistics show how serious a particular problem is.

Example 8.1

Specifications for a particular dimension of the Aft-Tech widget are .75 ± .04 centimeters. This means the designer believes that the widget will perform best and meet customer requirements when this dimension is between .71 and .79 centimeters. The process used to create the dimension has been subjected to a control chart study and is known to have a mean, $\bar{\bar{x}}$, of .75. The control limits for individuals are LNTL = .72 and UNTL = .78 (lower natural tolerance limit and upper natural tolerance limit, respectively). Does Aft-Tech have a problem?

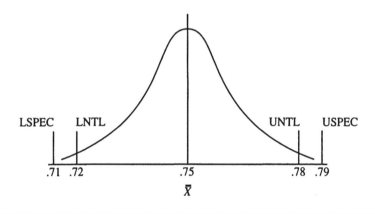

Figure 8.1   Relationship between Specifications and Natural Tolerance Limits

Solution

By the definition of three-sigma limits for individuals, $\sigma' = .01$. Since the specification, .71 and .79, are outside the limits that establish the process capability, it would be expected that virtually all of the material produced would be acceptable. This situation can be depicted graphically by drawing a normal curve to approximate the process and then superimposing the specifications, as in Figure 8.1.

Example 8.2

Specifications for weight on Aft-Tech's widget are 46 and 52 grams. Analysis of control chart data for the casting process indicates an average weight of 49 grams and a standard deviation of 1.1 grams. The natural tolerance limits, or the process capability, are as follows:

$$\text{UNTL} = 49 + (3)(1.1) = 52.3 \text{ grams}$$
$$\text{LNTL} = 49 - (3)(1.1) = 45.7 \text{ grams}$$

Does Aft-Tech have a potential problem?

Solution

In this instance, the process capability exceeds the specification limits. Specifically, the lower limit of 46 grams is larger than the lower process capability limit, and the upper specification, 52, is smaller than the upper process capability limit. The expected percentage of defective material is shown as the shaded areas under the normal curve in Figure 8.2. To determine the percentage of material that will not meet specification, z transform values must be calculated for both the lower and the upper specifications. For the lower spec,

$$z = \left(\text{spec} - \text{mean}\right)/\text{standard deviation} = \left(\text{spec} - \overline{\overline{x}}\right)/\sigma$$

$$= \left(46 - 49\right)/1.1 = -2.73$$

In the normal probability table, Table C of the Appendix, this z value corresponds to a probability of .4968 from the point to the mean, or .0032 from the point to infinity. On the low side, .0032, or .32 percent will not conform with the minimum weight specification.

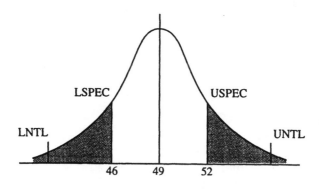

Figure 8.2   Relationship of Specs and Control Limits

For the upper specification, the same type of calculation is performed.

$$z = (52 - 49)/1.1 = +2.73$$

The corresponding probability from the upper spec to infinity is also .0032, or .32 percent. Thus a total of .64 percent of the material, .0032 + .0032, would be expected to not meet the customer requirements. If this percentage is considered unreasonably high, then Aft-Tech might take steps to change the process, change the specifications, verify customer requirements, or design an inspection plan to find as many of these defects as possible. Aft-Tech would also be able to assign a cost to producing this material outside of specifications. Some of the costs would include material scrapped, additional operations, and inspections.

Example 8.3

The Aft-Tech shaft works produces one shaft for which a certain specification is .84 ± .06 meter. The process has been in statistical control, with control limits for averages at .89 and .77 meters. The shafts, according to the in-process inspection records, are generally inspected in samples of size five. Aft-Tech's shaft management wants to know what percentage, if any, of the Aft-Tech shafts will not be produced to specifications.

Solution

In order to answer this question, the limits that define the process capability must be established. This information is available from the control limits for averages, so no additional sample inspections will have to be performed.

The first statistic to be derived from the given information is the process average, x. This is the midpoint of the control limits for averages, and may be obtained by solving the control limits' defining equations as two simultaneous equations with two unknowns:

$$UCL_{\bar{x}} = \bar{\bar{x}} + A_2\bar{R} = .89$$
$$LCL_{\bar{x}} = \bar{\bar{x}} - A_2\bar{R} = .77$$

Since A_2 is known to equal .58 when the sample size is five, the unknowns, $\bar{\bar{x}}$ and $\bar{R}$ can be determined.

$$.89 = \bar{\bar{x}} + (.58)\bar{R}$$

$$.77 = \bar{\bar{x}} - (.58)\bar{R}$$

The equations are added together to solve for $\bar{\bar{x}}$:

$$1.66 = 2\bar{\bar{x}}$$

$$\bar{\bar{x}} = .83$$

Returning to the original equations,

$$.89 = .83 + (.58)\bar{R}$$

$$\bar{R} = .103$$

Once the average range is known, the estimate of the standard deviation, s, can be calculated:

$$\sigma' = \bar{R}/d_2 = .103/2.326 = .044$$

The natural tolerance limits, which define the process capability, can now be calculated:

$$\text{UNTL} = .83 + (3)(.044) = .962$$
$$\text{LNTL} = .83 - (3)(.044) = .698$$

The upper specification for this example is .90 and the lower limit is .78. These fall well within the limits of process capability, and thus some defective material is to be expected. The expected percentage of defective material is shown as the shaded area under the normal curve in Figure 8.3. The z transform value for the lower specification, labeled area 1 in Figure 8.3, is

$$z = (.78 - .83)/.044 = -1.14$$

The corresponding portion of the areas below the lower specification is .127 or 12.7 percent.

The area above the upper specification, designated as area 2 in Figure 8.3, is also found by using the z transform.

$$z = (.90 - .83)/.044 = 1.59$$

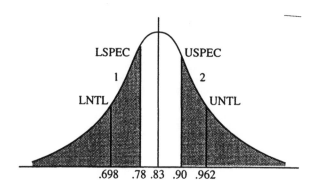

Figure 8.3   Relationship of Specs and Process Capability

The corresponding portion of values larger than the upper specification is .056, or 5.6 percent. The total percentage outside the specifications is 12.7 percent plus 5.6 percent, or 18.3 percent.

These values indicate a need for a change. Perhaps a different process could be selected, perhaps a modification to the existing process can be made. The desired effect would be to reduce the standard deviation. If this is not feasible, other alternatives include recentering the process at the midpoint of the specification limits. Sometimes a rather simple mechanical adjustment to a machine can cause a slight change in a dimension, reducing the percentage of material not meeting specifications.

In this example, it might be possible to center the process at .84, halfway between the spec limits of .78 and .90, instead of at the current value of .83. Moving the process up .01 meters would change the z transform values to

$$z = \pm\,(.90 - .84)/.044 = \pm\,1.36$$

This corresponds to a percentage of 8.69 outside of specs on both the high and low side, or a total of 17.38 percent defective material.

If the specifications cannot be changed, inspection procedures must be developed with the expectation of finding this level of defective material.

Example 8.4

A study of a certain cutoff operation in the Aft-Tech machine shop noted that pieces that were cut longer than the upper specification cost $.50 to rework, whereas parts that were cut shorter than the lower specification had to be scrapped at a cost of $1.50 per unit. The upper spec was 1.7 inches, and the lower spec was 1.3 inches. Control charts for the process indicated that it was presently centered between the specs with $\bar{\bar{x}} = 1.5$ inches; control limits for averages, based on samples of size three, were 1.68 and 1.32 inches. The control limits for individuals, which represented the process capability, were 1.81 and 1.19. This meant the process had a standard deviation of .104 inches. Consultation with the process engineer responsible revealed that the process *could* be adjusted or recentered to the nearest hundredth of an inch. Where should Aft-Tech recenter this process in order to minimize cost?

Solution

In this example, material that is below the lower specification costs three times as much as material that is above the upper specification, as reflected in the scrap cost to rework cost ratio of $1.50 to .50. Thus it would be logical to try to center the process so that three times as much material was outside of specifications on the high side as on the low side.

At present, the process is producing the same amount of defective material on the high end as on the low end. The z values are

$$z = \pm (1.70 - 1.50)/.103 = \pm 1.94$$

This z value corresponds to .0262 or 2.62 percent out of specifications on each side. Assuming material is produced in lots of 1,000, the inspection cost is the cost of producing and scrapping $(1,000)(.0262) = 26.2$ units, plus the cost of producing and reworking an additional 26.2 units. When an equal amount of material is outside of specifications on both the high and the low side, total cost is calculated as follows:

$$(\$1.50)(26.2) = \$39.30$$

$$(\$0.50)(26.2) = \underline{13.10}$$

$$\text{Total} = \$52.40$$

In order to minimize cost, the process average should be moved to a larger value so that more parts are reworked to meet specifications and fewer are scrapped. When three times as many parts have to be reworked as scrapped, the cost should be minimized.

Identifying the new value of the process average is a trial and error procedure. If x is moved from 1.50 to 1.60, the z values and corresponding percentages change as follows:

For the low side:

$$z = (1.30 - 1.60)/.103 = -2.91$$

This corresponds to a value of .0018, or .18 percent, for the portion of material outside of specifications on the low end.

For the high side:

$$z = (1.70 - 1.60)/.103 = +.97$$

This corresponds to a value of .1660, or 16.6 percent, for the portion of the material not meeting specifications. The ratio of .1660 to .0018 is 92.22, meaning that when the process is centered at 1.60 inches, over 92 times as much material is outside of specs on the high side as on the low side. For this case, the cost for the typical lot of 1,000 is $85.70. Based upon this it appears that the average should be lower, so that the ratio of high percentage outside the specifications to low percentage outside of specs is as close to three as the process will permit.

A value of 1.52 is the next selection. (It is wise to attempt to bracket the optimum point.) The z values are calculated.

On the low side:

$$z = (1.30 - 1.52)/.103 = -2.14$$

Out of specs = .0162

On the high side:

$$z = (1.70 - 1.52)/.103 = 1.75$$

Out of specs = .0401

The ratio of high out-of-specs to low out-of-specs is .0401/.0162 = 2.48. This is just under the desired ratio of 3. When this is the case the cost for a lot of 1,000 would be $44.35.

The next value tried is a mean of 1.53. Similar calculations give values of .0129 on the low side and .0495 on the high side. The ratio is 3.84. The cost for out of spec material is $44.10. If this process must be used and if the specifications cannot be changed, the process should be adjusted to aim at an average cutoff length of 1.53. This will minimize the combined cost of scrap and rework.

Process Capability Indices for Variables Data

A common desire of many control chart users is to be able to state a process's ability to meet specifications in one summary statistic (Kane, 1986, pp. 41–52). Such statistics are available and are called *process capability indices*. These indices are used to summarize internal processes as well as vendor processes.

Assumptions

All of the process capability indices we will discuss require three basic assumptions:

- Process stability
- Variables data
- Normality of process characteristic under study

It is common practice to assume that process capability indices are based on processes that are distributed normally. This ensures that all users of the capability index are "playing from the same score card."

Indices

Four process capability indices are commonly used: C_p, CPU, CPL, and C_{pk}. These indices are compared in Figure 8.4.

C_p

The C_p index is used to summarize a process's ability to meet two sided specification limits. C_p is computed as:

$$C_p = (USL - LSL)/6\sigma$$

Index	Estimation Equation	Purpose	Assumptions about Process
C_p	$\dfrac{USL - LSL}{6\sigma}$	Summarize process potential to meet two-sided specification limits	1. Stable process 2. Normally distributed process 3. Variables data 4. Centered process (process average equals nominal)
CPU	$\dfrac{USL - \bar{X}}{3\sigma}$	Summarize process potential to meet only a one-sided upper specification limit	1. Stable process 2. Normally distributed process 3. Variables data
CPL	$\dfrac{LSL - \bar{X}}{3\sigma}$	Summarize process potential to meet only a one-sided lower specification limit	(same as CPU)
C_{pk}	$C_p - \dfrac{\lvert m - \bar{X}\rvert}{3\sigma}$ where m = nominal value of the specification	1. Summarize process potential to meet two-sided lower specification limits 2. $\lvert m - \bar{X}\rvert/3\sigma$ is a penalty factor for the process's being off nominal. It is stated in terms of the number of natural limit units the process is off nominal	(same as CPU)

Figure 8.4   Process Capability Indices

Recall that a process's capability is defined to be the range in which almost all of the output will fall; usually, this is described as plus or minus three standard deviations from the process's mean, or within an interval of six standard deviations. Consequently, if a process's USL = UNTL = $\bar{\bar{x}} + 3\sigma$ and its LSL = LNTL = $\bar{\bar{x}} - 3\sigma$, the process's capability is 1.0.

A process capability of 1.0 indicates that a process will generate approximately three out of specification units in 1000, given the assumptions stated above. For centered processes, given the assumptions on the following page, there is a relationship between C_p, USL and LSL, and σ. These relationships are shown in Figure 8.5. Figure 8.5(a) shows a process with C_p = 1.0. This

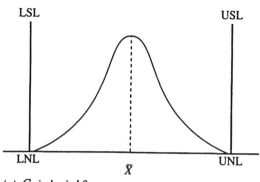

$$C_p = \frac{USL - LSL}{6\sigma} = 1.0$$

99.7% of output will be in-spec
$\sigma = (1/6)(USL - LSL)$

(a) C_p index is 1.0

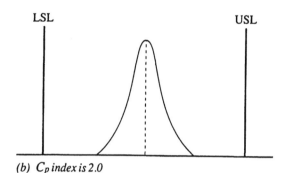

$$C_p = \frac{USL - LSL}{6\sigma} = 2.00$$

~100% of output will be in-spec
$\sigma = (1/12)(USL - LSL)$

(b) C_p index is 2.0

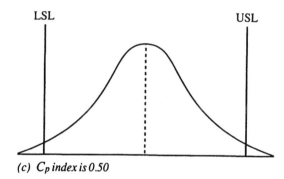

$$C_p = \frac{USL - LSL}{6\sigma} = 0.50$$

86.6% of output will be in-spec
$\sigma = (1/3)(USL - LSL)$

(c) C_p index is 0.50

Figure 8.5 Process Capability Indices

indicates that UNTL = USL and LNTL = LSL, hence 99.73 percent of the process's output will be within specification limits. We can say that if a process has a C_p = 1.0, the process standard deviation is one sixth of the distance between the upper and lower specifications limits.

Figure 8.5(b) shows a process with a C_p = 2.0. This indicates that the UNTL is halfway between nominal and the USL and that the LNTL is halfway between the LSL and nominal; hence, more than 99.999 percent of the process output will be within specifications.

For this situation

$$C_p = (USL - LSL)/6\sigma = 2.0$$

then

$$\sigma = (USL - LSL)/12$$

Thus we can say that if a process has a C_p = 2.0 the process standard deviation is one twelfth of the distance between the upper and lower specification limits. We must be careful not to infer that, because this C_p is twice the first instance, that this process is twice as capable. The assumption of normality includes the recognition that the normal curve is not a linear function.

Figure 8.5(c) shows a process with a C_p =0.5. This indicates that the USL is halfway between nominal and UNTL and that the LSL is halfway between the LNTL and nominal. Approximately 86.6 percent of the process output will be within specification limits.

CPU

The CPU index is used to summarize a process's ability to meet a one-sided upper specification limit. In many situations process owners are concerned only that a process not exceed an upper specification limit. For example, for products that can warp in only one direction, there is no LSL for warpage; the lower the warpage the better. However, there is a USL for warpage, the value for warpage that will critically impair the product's ability to meet customer needs. It can also be used in situations where we want to examine one side of a two-sided specification limit.

CPU is computed as

$$CPU = \left(USL - \bar{\bar{x}}\right)\big/3\sigma$$

The CPU index measures how far the process average, $\bar{\bar{x}}$, is from the upper specification limit in terms of natural tolerance limits. Natural tolerance limits, when added and subtracted from the process mean, $\bar{\bar{x}}$, yield the range in which a process is capable of operating — the process capability.

If a process USL = UNTL = $\bar{\bar{x}} + 3\sigma$, the CPU is 1.0. A CPU of 1.0 indicates that a process will generate approximately one and one-half out of specification units in 1000, assuming the process output is stable and distributed normally.

If a process's UNTL is greater than the USL, the CPU is less than one.

To determine the fraction of process output that will be outside of specifications for a given value of CPU, we use the normal curve relationship:

$$z = \left(USL - \bar{\bar{x}}\right)/\sigma$$

CPL

The CPL index is used to summarize a process's ability to meet a one-sided lower specification limit. It is, in effect, the mirror image of the UPL. It can also be used in situations where we want to examine one side of a two-sided specification limit.

CPL is computed as

$$CPL = \left(\bar{\bar{x}} - USL\right)/3\sigma$$

The CPL index measures how far the process average, $\bar{\bar{x}}$, is from the lower specification limit in terms of natural tolerance limits.

If a process LSL = LNTL = $\bar{\bar{x}} - 3\sigma$, the CPL is 1.0. A CPL of 1.0 indicates that a process will generate approximately one and one-half out-of-specification units in 1000, assuming the process output is stable and distributed normally.

If a process's LNTL is less than the LSL, the CPL is less than one.

To determine the fraction of process output that will be outside of specifications for a given value of CPL, we use the normal curve relationship:

$$z = \left(\bar{\bar{x}} - LSL\right)/\sigma$$

C_{pk}

The C_{pk} index is used to summarize a process's ability to meet two-sided specification limits while not necessarily being centered between those limits. The C_{pk} is calculated as the smaller of the following:

$$\left(USL - \overline{\overline{x}}\right)/3\sigma \quad \text{and} \quad \left(\overline{\overline{x}} - LSL\right)/3\sigma$$

When C_{pk} is less than 1.0 it indicates that the process is not capable of meeting specification limits. When C_{pk} is equal to 1.0 it is just capable of meeting specifications, and when C_{pk} is greater than 1.0 then we say the process is "more" than capable of meeting requirements.

If a firm is pursuing continuous improvement it will be striving to move the capability indices towards infinity. A firm that exists in a defect detection mode will not know the process capability indices for its various processes. On the other hand, a firm operating in a defect prevention mode will know the values for its various processes and will be striving initially for the values to be equal to 1, and then to exceed 1.

Limitations of Capability Indices

Several potential problems exist when using the C_p and C_{pk} indices. First, if a process is not stable, C_p and C_{pk} are meaningless statistics. Second, not all processes meet the assumption of normality. Hence, the naive user of capability indices will incorrectly assess the fraction of process output that will be outside of specifications. Experience shows that naive users of the indices frequently confuse C_p and C_{pk}; they think they yield the same information about a process. Lastly, some users assume a linear relationship between the C_p and C_{pk} indices and the amount of material that will be outside of specifications.

Example 8.5

Each of the process capability indices discussed earlier in this chapter is calculated using the following data from a camshaft manufacturing operation. For this camshaft the average case hardness depth is 4.43 mm, the average range for case hardness depth is 1.6 mm, the upper specification is 10.5 mm, and the lower specification limit is 3.5 mm. (See Figure 8.6.) Data for control charts were collected using a subgroup size of n = 5.

Given the above information calculate C_p, CPU, CPL, and C_{pk}.

Solution

The initial step is to calculate the estimate of the process standard deviation. Using the relationship $\sigma' = R/d_2$, the standard deviation is estimated to be $1.6/2.326 = .688$ mm. The calculation of C_p is shown

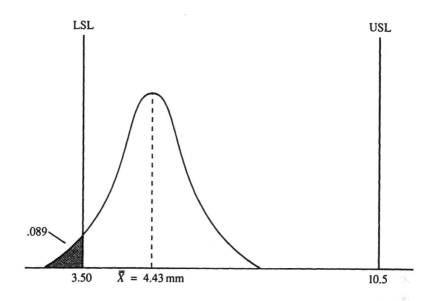

Figure 8.6   Fraction of Camshafts Outside of Specification

$$C_p = (USL - LSL)/6\sigma = (10.5 - 3.5)/(6)(.688) = 1.70$$

This C_p indicates an extremely capable process that will almost never produce outside-of-specification product. Obviously, this is completely false; since a calculation of the LNTL = x – 3σ′ shows that LNTL is 2.37 which is lower than the LSL. *The C_p index failed to accurately state the process capability because of the C_p's assumption that the process is centered between the specification limits.*

The CPU calculation is shown below.

$$CPU = \left(USL - \bar{\bar{x}}\right)/3\sigma' = \left(10.5 - 4.43\right)/\left(3\right)\left(.688\right) = 2.94$$

The CPU accurately indicates that the process is operating well within the USL of 10.5 mm.

Computing the CPL gives the following result:

$$CPL = \left(\bar{\bar{x}} - LSL\right)/3\sigma' = \left(3.5 - 4.43\right)/\left(3\right)\left(.688\right) = .45$$

The CPL accurately indicates that the process is not capable regarding the LSL. If desired, we can calculate the expected percentage of camshafts that will be produced outside of this specification limit via the following calculations:

$$z = \left(LSL - \bar{\bar{x}}\right)\big/\sigma' = \left(3.5 - 4.43\right)\big/.688 = -1.35$$

From Table C we see that 8.9 percent of the camshafts would be expected to be out of specification on the low side.

The C_{pk} is calculated as the smaller of the following:

$$\left(USL - \bar{\bar{x}}\right)\big/3\sigma \quad \text{and} \quad \left(\bar{\bar{x}} - LSL\right)\big/3\sigma$$

These correspond to the just calculated CPU and CPL of 2.94 and .45, respectively. This C_{pk} indicates that the process will produce a good deal of defective product.

Summary

Process capability tells the analyst within what limits the process is capable of producing. This information makes possible identification of those processes, as represented by machines, material, and people, that simply are not capable of meeting specifications. If the machines cannot be changed, the people better trained, the specifications modified, or the process recentered, then the organization must learn to live with a certain percentage of its product not conforming to the desired specification.

Perhaps the best way to state process capability is through the statement of the expected percentage not conforming. C_p and C_{pk} can potentially cause more problems than they can provide benefits.

Reference

Kane, V., "Process Capability Indices," *Journal of Quality Technology*, January, 1986, pp. 41–52.

Practice Problems

1. A manufacturer of ambi-helical hexnuts has established a torque specification of 42 inch pounds minimum. Control limits for averages are 54 and 40. The following values were noted in the inspection records for the past two samples:

   n1: 43, 46, 48, 50, 53, 51, 44, 50
   n2: 48, 50, 49, 49, 50, 48, 50, 49

   (a) Is the process in control?
   (b) What percentage of material is not being produced to specifications?

2. Forty samples of three were discovered to have an upper control limit for ranges of 9. If it is known that the process average is 50, what is the process capability?

3. Specifications for bilateral blivets are 11.9 ± .75 meters. Twenty samples of three had an average of 12.5 and an average range of 1.8 meters. How many bilateral blivets in a sample of 200 would be expected to meet specifications?

4. Fifty samples of four have an average range of 10.295 and an average average of 24. If specifications indicate a maximum allowable tolerance of 24 ± 8, how many items in a sample of 600 would be expected to not meet the specifications?

5. The specified bend on a formed part is 40 degrees ± .3 degrees. Based on samples of size six, a control chart analysis indicated that there was an observed average of 40.2 degrees with an average range of .4 degree. How many parts in a lot of 14,400 would not be produced to specifications?

6. A process has an upper specification of 85 and a lower specification of 49. Control limits, based on samples of size seven, are 80 and 55 for averages. What percentage of parts would be expected to be outside of specifications?

7. A process has a mean of .183 and an average range of .06. The sample size is six. Specifications are .185 ± .025. What percentage of material would be expected to be outside of specifications? If the process could be recentered, what would be the minimum percentage defective expected?

8. A process for testing thread strength indicates that control limits for averages are 22 ksi and 47 ksi. The specifications require a minimum strength of 10 ksi. The control limits were established using a sample size of five. What percentage of thread would be expected to be outside of specifications?

9. Specifications on a part are 182 and 187. In the past the process has averaged 184.7. Averages of four samples of size four were measured to be 183, 186, 187, and 185. The average range in the past has been 10.295. What is the best estimate of process capability?

10. A process has an average of 72 and a standard deviation of 7.6. What is the process capability?

11. For a certain manufacturing process, the cost of repairing parts that are below specs is $1.25 per unit. The parts above specs cost $5.00 to repair. The lower spec is .85 and the upper spec is 1.15. Control limits for averages, based on samples of size eight, are .89 and 1.10. Where should the process average be recentered in order to:
(a) Minimize the percentage of defective material produced?
(b) Minimize the cost of the defective material produced?

12. Control limits for a process were set using samples of size two. These limits were UCLx = .406 and LCLx = .328. Specifications for this process were .310 and .420. What percentage of defective material would this process be expected to produce?
Determine the appropriate capability indices for the situations described in problems 13 and 14.

13. A contact lens has a specification requirement for center thickness of .22 microns ±.15 microns. A process capability study measured 100 individual lenses under normal operating conditions. It found the average was .22 and the standard deviation was .04 microns.

14. The manufacturer's gap on a corrugated box is supposed to be .375 inches ±.125 inches. A preliminary study indicated that, based on 50 samples of 5, the process was stable with control limits for averages at .22 and .51. Control limits for ranges were .5275 and 0.

15. In problem 14, how many boxes in an order of 1,000 would we expect to find outside of specifications?

16. A data entry operation has averaged 25 errors per shift per operator. This has had a standard deviation of 4 errors.

(a) If spec limits were 13 and 37, what would the percentage outside of specs be?

(b) If the standard deviation were reduced, through process improvement, to 2, what would the percentage outside of specs be?

17. Recent literature has spoken of six-sigma quality. This means that spec limits instead of being at the $\pm 3\sigma$ limits are targeted to be at the $\pm 6\sigma$ limits. Comment on what six-sigma quality means in terms of defects per million units produced or operations performed.

9 SPC for Continuous Quality Improvement

Introduction

This book is primarily written to explain the fundamentals of statistical quality control. SQC is concerned with the application of proven statistical methods to manufacturing and service processes. Often SQC and SPC are used interchangeably. This is not a good practice.

SPC, which stands for statistical process control, is the application of statistical methods to monitor and adjust processes. It focuses on identifying when process problems or abnormalities occur and on finding ways to eliminate or prevent these and improving the performance of the processes.

SPC is a procedure that helps organizations to improve quality. Many organizations incorrectly give this name to what could better be called a total quality improvement process. SPC helps organizations improve quality by solving problems through the identification and measurement of critical quality characteristics. It requires the use of the knowledge of the individuals closest to and most familiar with the process.

As such, SPC has three major components. First, it helps to identify process quality problems. Second, it helps to identify the causes of those problems, Third, it helps to evaluate the effect of process changes on the organization.

SPC is a method of mobilizing the abilities of all members of the organization to improve the quality of the processes that produce the goods and services it provides.

SPC Model

SPC is best defined as part of the model and method for total quality improvement. SPC, when used in this context, can be modeled and used as a guide for overall quality improvement.

Quality improvement begins with management commitment. For SPC and total quality to work, all members of management must demonstrate commitment to improving the processes of the organization. The demonstration of this commitment is a difficult and involved process. Organizations are typically driven by production schedules, production costs, and quality. The key to making SPC and total quality improvement work is to give quality at least the same level of emphasis as maintaining schedules and keeping costs as low as possible. This is often difficult.

All of us are constantly sending signals. Members of management can go on record as saying that quality is important. They can issue quality statements, vision statements, and mission statements; but they *can also send signals that if the product is not delivered on time all hell will break loose.* When that happens the wrong signals are received and quality suffers.

Commitment is the key for continuous improvement or SPC to work. The point that has to be remembered is that everyone sends signals. Our actions speak louder than our words. Most important is what might be called the Pogo effect. This is based on the famous cartoon in which the line, "We have met the enemy and he are us," is uttered. In terms of management commitment, the Pogo effect means that, to someone else in the organization, each of us is a member of management. Our actions carry far more weight than any words we may ever state.

Following the statement and demonstration of commitment by management to quality improvement, SPC involves several mechanical steps. The first of these is called process identification. Every activity of the organization must be viewed as a process. A process has inputs, which come from suppliers, a transformation, which adds value to the inputs, and outputs, which are given to the customers of the process, as shown in Figure 9.1. In addition, there is feedback from the customers to the suppliers and value adders of the process. Members of the organization must begin to think of the customers and suppliers as being within or internal to the organization as well as external to the organization. The operation of the organization is, in effect, a series of interconnected mini-processes which, when linked, will combine to provide the product for which the organization wishes to be known.

Figure 9.1 Basic Process

Each process has measurable inputs and outputs. The output for one process is the input for the next. This holds true whether it is internal to the organization, as shown in Figure 9.2, or external to the organization. Each of us is both a customer and a suppler for other processes in the organization.

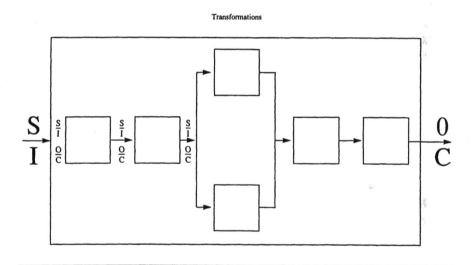

Figure 9.2 Interconnected Processes

SPC uses this process identification as the basis for process control and process improvement. After the process is defined, the critical characteristics are defined. The process of identifying the characteristics of the process that are critical to the overall consistency of the product or service produced is the next step in applying SPC. Relatively few process characteristics are usually responsible for most of the quality problems that the process faces. In addition, most of those characteristics are inherent or indigenous to the process. The process will not be improved until these characteristics are addressed by the people, meaning management, who are committed to improving the processes. Relatively few characteristics can be addressed by the people closest to the process. Once critical characteristics are identified, SPC uses process monitoring.

Process monitoring involves the use of control charts. In SPC, as in any application, control charts identify the nature and type of variation present in the process characteristic being monitored. Normal variation is the variability that is always present in a stable process; abnormal variation is unusual and is due to special causes. Normal variation is reduced only by changing the process; abnormal variation is reduced by identifying and eliminating the special causes of the abnormal variation.

The SPC continuous improvement model is based on the concept that consistency at the target value represents quality. In statistical terms, the ideal is for each process mean to be the target or nominal value with a standard deviation of zero. Process monitoring tells us when we are at that point or where we are relative to that goal. Process improvement suggests what we might do to get closer to the objective. We use information about process performance to take action to improve the process.

Once the process is improved, we have, in effect, a new process, which must be redefined. The new critical characteristics must be identified, their variation monitored, and the process controlled when abnormal or special causes of variation enter the process. Consistent with the philosophy of continuous process improvement, we must once again improve the process.

SPC sets up a continuous cycle of ongoing improvement, which continues as long as management commitment is present. It is as if we are moving up a long, never ending hill. Management commitment provides the energy for continuing the process. While the commitment is present, the progress will continue. Once the commitment ends, the progress will continue for a short period of time, but it will surely cease. And once the source of the momentum disappears, the body in motion will come to rest. Since quality improvement is an uphill journey, without the upward motion, the organization will rapidly begin to go downhill.

SPC Tools

When SPC is used synonymously with total quality improvement, it uses some familiar and some not so familiar tools for identifying critical process variables. This section briefly describes some of the SPC tools that are commonly used. These tools have, in some contexts, picked up the designation, Simple Powerful Charts. Some have been described in great detail in this book; others may require the reader to seek further definition in other texts.

Check Sheets

Check sheets are used to quantify the relative appearance of specific quality problems. Check sheets include information about the specific definition of problems, identification of when those problem occur, measurement of the number of occurrences of those problems, and a tabulation of the frequency with which those problems occur. The check sheet is a basic data gathering tool which provides information regarding which quality and process problems are actually occurring.

Pareto Analysis

Pareto analysis differentiates the significant few from the trivial many. Based on the principle that a few quality problems cause most of the "damage", the principle guides the strategy of investing resources toward the goal of eliminating the causes that are responsible for either most of the problems or most of the cost of the problems. A sample Pareto chart is shown in Figure 9.3. In the case of this sample, it shows that almost 30 percent of an apparel manufacturing company's problems are due to mismatched pieces of material.

Cause-and-Effect Diagram

This is a tool which is used to guide brainstorming. Using the quality problem identified via the check sheet and Pareto analysis, the cause-and-effect diagram attempts to identify potential causes of the quality problem. By using a diagram like the one shown in Figure 9.4, major causes, contributing causes, and contributing causes for the contributing causes are identified. Using major categories, such as materials, machinery, and work methods, the input of those closest to the process can be identified and sorted, often providing insight as to what the critical process characteristics are.

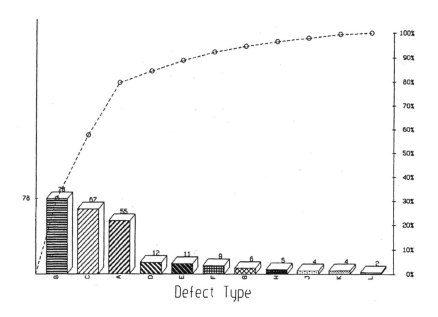

Figure 9.3   Pareto Chart

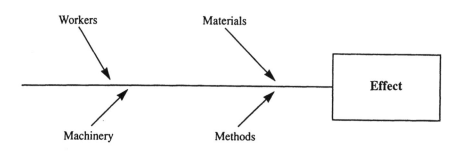

Figure 9.4   Cause-and-Effect Diagram

Basic Statistics

Basic statistical measures, such as histograms and descriptive statistics are valuable continuous improvement tools. Histograms intuitively show where the process is centered, how much it varies, and the general shape of the distribution. A process centered near the target of nominal specification value can be seen in the histogram, as can one that consistently runs high or low relative to nominal. The spread can be compared with tolerance or spec limits. Too many values outside of the specs can lead us to the conclusion that the process is incapable. If the histogram shows the values to be centered near the middle of the distribution, with more values near the center than at the tails, and with a gradual decrease of values as the tails of the distribution are approached, then the sample can be said to be normally distributed. Normality allows us to predict limits of normal process variations and helps us to identify when abnormal variation enters the process. All three of these characteristics tell the trained observer how the process is working. Histograms provide a natural lead into the use of descriptive statistics.

Descriptive statistics quantify the intuitive information that the graphical histogram shows. The mean tells us where the process is centered and the range or standard deviation gives a measurement of the amount of variation present in the process.

Control Charts

Control charts show control limits and a picture of how a process varies over time. Control limits show the normal variation for the process being charted. When only normal variation is present, the process is stable. Stable processes are good candidates for improvement.

Summary

SPC is the application of statistical concepts to process monitoring and control. As it is used in most businesses, it is synonymous with total quality or continuous quality improvement. The underlying principle is the one of *preventing* problems rather than dealing with the results of the problems after they occur. When problems do occur SPC is useful in helping to determine the cause(s) of the problems. The motto of SPC is prevention, not detection.

Acceptance Sampling

Introduction

> Even before the 1920's, industry was learning how to do things more scientifically, using such techniques as the Gantt chart, under principles developed by Taylor, Gilbreth, and others in a movement called "scientific management." And in 1922, G.S. Radford published his book, *The Control of Quality in Manufacturing.* But the Bell System group was organized with what proved to be a new approach to the problem of quality, having as its broad objective the quality of performance of a rapidly growing nationwide communications system. New techniques and procedures were sought for getting the highly needed, highly uniform quality of the many elements of the complete telephone system. At that time, when it was common practice for products and services to strike a balance between schedules and costs, it was proposed to add a third factor — quality ...A number of techniques, such as ...acceptance sampling inspections plans, were essentially statistical in nature. (Dodge, 1969, p. 78)

Thus, acceptance sampling has been used in some form since the 1920s. When a manufacturing concern purchases parts, when a restaurant purchases food, or when any other organization purchases raw material for further processing, it hopes to purchase only material meeting specifications. But since no supplier ever provides completely acceptable material all of the time — random variation enters every process — most concerns inspect the product that has been ordered to determine if the material conforms to the agreed upon quality standards.

Inspection is the process of comparing actual measurable characteristics with predetermined standard characteristics.

There are three basic ways to perform inspections. First, every single part that is received may be inspected. This is referred to as 100 percent inspection. It has several drawbacks. Often the cost is prohibitive, as is the time required to inspect every piece of material received. Also, Samson, Hart, and Rubin have vividly illustrated that because 100 percent inspection often involves long periods of highly repetitive and monotonous work, it is not always 100 percent effective (Samson et al., 1970, p. 72). In cases in which testing destroys the product being tested, 100 percent inspection is of course not even an option.

On the other side of the fence, is 0 percent inspection. The organization accepts the parts on faith, believing that they must all be of an acceptable quality level because the supplier says so. It would indeed be a beautiful world if everyone could operate entirely on faith and all business deals were aboveboard. It would also be wonderful if the normal random variation that is always present could invariably be contained to acceptable amounts.

A third inspection option is a procedure known as sampling inspection. There are several types of sampling procedures. The chief types are described in Table 10.1.

Table 10.1   Types of Sampling

Type	Description
Random sampling	Selecting items so that any of the remaining items in the population has an equal likelihood of being selected.
Stratified sampling	Selecting items that come from known subgroups so that the sample is proportional to the distribution of the original subgroups.
Systematic sampling	Selecting items at defined period intervals.
Quota sampling	Selecting items based upon the known distribution of a number of factors, such as time of day, department, or machine location.
Judgment sampling	Selecting samples based entirely on the judgment of the sampler.
Sequential sampling	Sampling inspection in which, after each unit has been inspected, the decision is made to accept the lot, not to accept the lot, or to inspect another sample.
Chain sampling	Sampling inspection in which the criteria for acceptance and rejection of the lot depends on the results of the inspection of immediately preceding lots.

Table 10.1   Types of Sampling (continued)

Type	Description
Multiple sampling	Sampling inspection in which, after each sample is inspected, the decision is made to accept a lot; not to accept it; or to take another sample to reach the decision. There may be a prescribed maximum number of samples, after which a decision to accept or reject the lot must be reached.
Skip lot sampling	In acceptance sampling, a plan in which some lots in a series are accepted without inspection when the sampling results for a stated number of immediately preceding lots meet stated criteria.
Single-level continuous sampling	Sampling inspection of consecutively produced units in which a fixed sampling rate is alternated with 100 percent inspection depending on the quality of the observed sample.
Multi-level continuous sampling	Sampling inspection of consecutively produced units in which two or more sampling rates are alternated with 100 percent inspection, or each other, depending on the quality of the observed product.
Single sampling	Sampling inspection in which the decision to accept or not to accept a lot is based on the inspection of a single sample of size n.
Double sampling	Sampling inspection in which the inspection of the first sample of size n_1 leads to a decision to accept a lot; not to accept it; or to take a second sample of size n_2, and the inspection of the second sample then leads to acceptance decision regarding the lot.

Above definitions adapted from *Glossary and Tables for Statistical Quality Control,* American Society for Quality Control, Milwaukee, Wisconsin, 1983. Reprinted with permission.

Each type of sampling procedure is particularly suitable for certain types of information gathering. Only random sampling, systematic selection, and stratified sampling would normally be used for acceptance sampling. "The reliability of the other types depends too greatly on knowing a good deal about the particular lot being sampled. The large sums of money generally involved in the decision of acceptance sampling warrant a sampling with high reliability." (Samson et al., 1970, p. 75) The most frequently used type of sampling is random sampling. Ideally, random sampling should be part of most sampling inspection plans in a manufacturing organization because it protects against errors due to unrecognized cycles or patterns.

There are several advantages to using sampling inspection rather than 100 percent inspection. First, the inspection can be performed more economically. Specifically, when sampling inspection is used:

- Less time is required for inspection.
- Fewer inspectors may be needed.
- Inspection is more of an audit and less of a police function.

Second, the precision of the inspection is increased. Since fewer parts are inspected for the same characteristics:

- The inspection can be more detailed and more quality characteristics can be inspected.
- Inspectors are likely to do a better job because there is less monotony.

Sampling inspection is not perfect; errors will be made. There is a certain amount of risk involved in using any sampling plan because all parts are not examined. However, this risk, which is present any time sampling inspection is performed, can be quantified when acceptance sampling is used.

As Table 10.1 indicates, there are several ways to sample, and even when random sampling is used, there are several different types of sampling plans that may be used. The basic type is a single sample sampling plan. This plan is the easiest to understand, but relatively large samples are required before decisions on acceptability can be made when material is of either very high or very low quality. In a single sample sampling plan, the inspector inspects a sample and, based on the evaluation of the quality, recommends either accepting or not accepting the material the sample was taken from.

Following are several terms that are used to describe single sample sampling plans.

The **sample size, n,** is the number of parts the inspector will inspect in a given batch or lot of material. These parts are selected at random from the entire batch or lot.

The **lot size, N,** is the number of items of a particular product either received in a single shipment or produced in a given production run.

It is the entire population of material that is submitted for acceptance at one time.

The **acceptance number, c,** is the maximum number of parts not meeting customer requirements or specifications that a sample may contain before the entire lot is no longer considered acceptable.

A **defective** is a part that fails to conform with quality standards and thus is classified as unacceptable.

When any kind of sampling plan is used the inspector, or operator if self-inspection is employed, must make a decision about the acceptability of the material being inspected. Under a single sample sampling plan, the inspector checks a single sample of size n and classifies the sample as acceptable if there are *c or fewer defects in the sample.* If there are more than c defects in the sample the sample is considered unacceptable and the entire lot is rejected.

Sampling Risks

Any sampling plan, whether single, double, or multiple, involves some risks. Obviously, when just a random sample of material is inspected, the inspector does not have an opportunity to check every part produced. Thus some material that should not be accepted will be. Also, the preconceptions of the inspectors may cause them to call some items defective, when really they are acceptable. Those individuals charged with the inspection function may feel that their jobs are not being done if not material is classified as being acceptable. After all, the job description for inspector usually includes the phrase, "to find defective material." This risk is present in 100 percent inspection as well as in all forms of sampling inspection.

Because sampling inspection is statistically based, these risks can be defined. It is possible to determine the chances of making mistakes before any action is taken. This might be compared with setting your own odds before entering a game of chance. And of course, when the organization sets its own odds, it is possible to set them in the "house's" favor by minimizing producer's risk.

Producer's risk is the probability of rejecting material that should be accepted. This is a type I statistical error and is designated with the symbol, α.

Consumer's risk is the probability of accepting material that should be rejected. This is a type II statistical error is designated with the symbol, β.

From these definitions we can see that the probability of accepting material that should be accepted is $1-\alpha$. Material that should be *accepted* and material that should be *rejected* are ambiguous phrases. In acceptance sampling plans, the quality of the material is usually described in terms of the percentage of material that does not meet quality standards, called the percentage defective.

The **percentage defective,** designated by the symbol p' is the ratio of the number of defects in the lot to the number of items in the lot.

These acceptable quality and the rejectable quality have two special names.

The **acceptable quality level,** or **AQL,** is the maximum percentage defective that is considered satisfactory as a process average.

The **lot tolerance percentage defective,** or **LTPD,** is the percentage, or proportion of defective items for which the consumer wishes the probability of acceptance to be restricted to a specified low value (ASQC, p. 52). Material at or below the LTDP level is generally considered to be unacceptable. It should be interpreted to be a rejectable quality level.

Since risks are involved in sampling, there are times when high quality material will be rejected or low quality material will be accepted. These situations have been defined as the producer's and consumer's risks, respectively. In general, the chance of accepting (or rejecting) any lot of material can be calculated if the percentage of defective material, p', is known.

The **probability of acceptance** of any given lot of material, designated **Pa,** is the chance that the sample inspected will be found to meet all quality standards. Probabilities of acceptance range from 0 to 1.

Operating Characteristic Curves

The interrelationships of the risks, probabilities, and qualities defined in the previous section can be shown graphically on an operating characteristic, or OC curve.

The **operating characteristic curve,** or OC curve, shows the relationship between lot quality and the likelihood of accepting the lot for a given sampling plan.

Some important features of the OC curve include the following:

- When normal material is perfect, or 0 percent defective, it should be accepted all the time. The OC curve shows that the probability of accepting material that is 100 percent acceptable is 1.0, or a sure thing.
- At the other extreme, when material is all bad, or 100 percent defective, there should be no chance of accepting it. The OC curve shows that the probability of accepting material that is 100 percent defective is 0.
- If the material is of acceptable quality, or AQL percent defective, there should be only a small chance of rejecting it. Therefore, Pa should be high. The OC curve shows that the probability of accepting material that should be accepted, the probability of accepting material that is AQL percent defective, is $1 - \alpha$.
- If the material is not generally considered acceptable, or of a quality that should be rejected, or that is LTPD percent defective, there should be only a small chance of accepting it. The OC curve shows that the probability of accepting material that is LTPD percent defective is β.

Figure 10.1 shows the general shape of a typical OC curve. The exact shape of a specific curve is dependent on the particular values of AQL, LTPD, α, and β. The curve will show the probability of accepting a lot using a specified sampling plan for any given value of p'.

The shape of the ideal OC curve is somewhat different from that of the curve in Figure 10.1. By definition, any lot that is AQL percent defective or better should be accepted all the time, and any lot that is more than AQL percent defective should never be accepted. Ideally, AQL should also be 0 percent defective. Figure 10.2 shows the idealized OC curve, in which this principle is carried to the logical extreme. Only material that is 0 percent defective is accepted. All other material is rejected.

Sampling Probabilities

To develop sampling plans that meet the desired α and β risks and maintain the desired AQL and LTPD quality levels, the appropriate statistic and

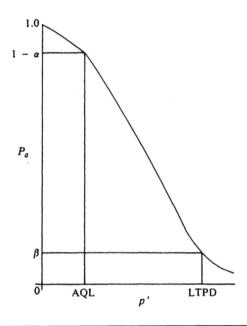

Figure 10.1   Typical OC Curve

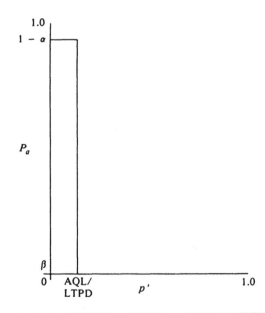

Figure 10.2   Ideal OC Curve

probability relationships must be used. The actual development of a sampling plan is best illustrated with the single sample sampling plan.

The objective in developing a sampling plan is to determine the probabilities of accepting lots of varying quality. What is the chance of accepting material that should be rejected? What is the chance of rejecting material that should be accepted? What is the chance of accepting material that should be accepted? The answers to these questions are found through development of an OC curve and/or the associated statistics.

Because sampling plans reflect the number of defects per sample, the Poisson probability distribution is typically used:

$$P(x) = \frac{(np')^x e^{-(np')}}{x!}$$

Example 10.1

Use the Poisson probability distribution to determine the probability of finding exactly three defects in a sample of 80, if the material is really 4 percent defective.

Solution

In this example

$$x = 3 \quad n = 80 \quad p' = .04$$

The probability is calculated using the equation

$$P(3) = \frac{[(80)(.04)]^3 e^{-(80)(.04)}}{3!}$$

$$= .2226$$

This is the probability of finding exactly three defects in a sample of 80 if the material is actually 4 percent defective.

The probability of accepting a lot depends on the sampling plan. If acceptance of the material in Example 10.1 was dependent on finding three or fewer defects in the sample of 80, the probability of accepting would be the probability of finding three or two or one or zero defects in the sample. In general, a single sample sampling plan suggests that material be accepted with c or fewer defects in the sample. In order to determine the probability of accepting the lot, the probability of finding exactly three defects, exactly two defects, exactly one defect, and exactly zero defects would have to be calculated. These would then be summed to determine the probability of accepting the lot.

Example 10.2

What is the probability of accepting a lot that is 5 percent defective using a single sample sampling plan in which n = 100 and c = 4?

Solution

In order for a lot to be accepted under this sampling plan, there would have to be four or fewer defects in the random sample of 100. The probability of acceptance is determined as follows:

$$Pa = P(0) + P(1) + P(2) + P(3) + P(4)$$

Use of the equation for the Poisson distribution could prove to be quite tedious. Instead, the cumulative Poisson distribution, shown in Table I of the Appendix is used. This table shows the probability of c or fewer defects for given values of np'. For the p' value specified in this example np' = (100)(.05) = 5.0. The corresponding value read from the table is .440 (at the np' = 5.0 row and c = 4 column intersection). With this sampling plan there is a 44 percent chance of accepting material that is 5 percent defective. If the percentage were changed, the probability of acceptance would change. An OC curve could be plotted after a sufficient number of calculations had been made.

Example 10.3

Determine the probability of accepting a lot for which it is believed p' = .08, using a single sample sampling plan in which n = 85 and c = 3.

Solution

In order for Table I to be used, the value of np' must be calculated:

$$np' = (85)(.08) = 6.8$$

The value in the c = 3 column and the np' = 6.8 row is .093. This is the probability of finding three or fewer defects in the sample. It is also the probability of accepting the lot if the material is 8 percent defective.

The OC curve for an acceptance sampling plan shows the general shape of the risks associated with a sampling plan as a function of the quality levels. The shape shows whether the consumer is fairly protected or the producer has the advantage. An OC curve like the one shown in Figure 10.3 gives the advantage to the consumer, since only material of very good quality will be accepted. Figure 10.4 shows a case where the producer has the upper hand — material with a very high percentage of defectives will be accepted much of the time. The general shape of the OC curve can provide valuable information about the sampling plan in question.

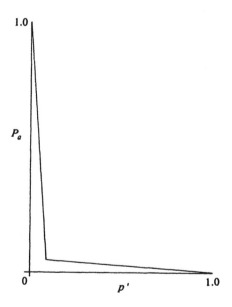

Figure 10.3   OC Curve Favoring Consumer

Plotting the OC curve for a single sample sampling plan is a relatively straightforward procedure. All that is needed is the sample size, the acceptance number, some specified values for the percentage defective, and the cumulative Poisson table.

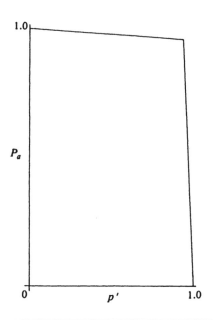

Figure 10.4   OC Curve Favoring Producer

Example 10.4

Graph the OC curve for the single sample sampling plan in which n = 150 and c = 5.

Solution

In order to graph any curve points are required. The particular points needed in this instance are values of p′ and the corresponding values of Pa.

Two points are intuitive and immediately obvious. When the material is 0 percent defective the chance of accepting lots with a sampling plan is 1.00. When the material is 100 percent defective, the probability of accepting will be 0.00. The intermediate points have to be calculated using the Poisson probability relationships just developed.

An initial value of p′ = 2 percent is arbitrarily selected. For np′ = (150)(.02) = 3.0 and c = 5. Table I gives a probability of .916. This is the likelihood of accepting a lot that is actually 2 percent defective. The same procedure is used for other values of p′ until enough values have been calculated to plot a good curve. Some additional calculations are tabulated in Table 10.2. Figure 10.5 shows the OC curve for this plan.

Table 10.2   Sample Points for an OC Curve

p′	np′	Pa
0.00	0.0	1.000
0.02	3.0	0.916
0.03	4.5	0.703
0.04	6.0	0.446
0.05	7.5	0.242
0.06	9.0	0.116
0.01	10.5	0.050
1.00	15.0	0.000

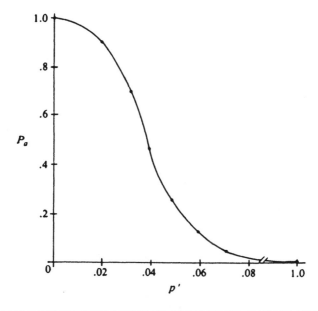

Figure 10.5   OC Curve for Example 10.4

OC curves show the probability of accepting material for all possible percentages of defective material. If α, β, AQL, and LTPD are known, the OC curve can be sketched. The more points available the better the curve will be.

Some situations call for developing a single sample sampling plan for specified values of α, β, AQL, and LTPD, rather than comparing the sampling plans based on the shape of the OC curve. The task is to determine values of n and c that *most closely* match the given values for α, β, AQL, and LTPD. This is a trial and error procedure and is best explained through examples.

Example 10.5

Develop a single sample sampling plan to fit the following conditions;

$$AQL = .02 \quad LTPD = .06 \quad \alpha = .10 \quad \beta = .10$$

Solution

To specify a sampling plan, n and c must be determined. Because of the way the Poisson table is constructed, the technique is to control c and to solve for n. The methodology is a trial and error procedure. The final values for n and c are the best approximations available. They come closest to meeting the specified risks at the desired quality levels.

Before any trials can be undertaken, two preliminary calculations must be made. First, the probability of accepting material that should be accepted, $1 - \alpha$, must be determined. In this example, $1 - \alpha$ is $1 - .1 = .90$. Second, the ratio of LTPD to AQL must be calculated. For this example the ratio is $.06/.02 = 3$.

Now the trial and error procedure can begin. For the first trial, a value of c = 1 will be arbitrarily specified. This value merely represents a starting point, depending on the results of this trial, this value for c may be accepted or a new one may be tried. In Table I the value corresponding to c = 1 that comes closest to the $1 - \alpha$ value is found. In this case, the closest value is between .910 and .894. Interpolation between the np' values of .5 and .55 shows the closest np' value to be .53. Since this value of np' corresponds to $1 - \alpha$ it has the special designation of nAQL. There is a $1 - \alpha$ chance of accepting material that is AQL percent defective.

Since LTPD was determined earlier to be three times as large as AQL, obviously nLTPD should be three times as large as nAQL. For this example

$$nLTPD = (3)(nAQL) = (3)(.53) = 1.59$$

The value of β corresponding to nLTPD = 1.59 and c = 1 is .528; this is the probability of accepting material that should be rejected. Comparing this value the desired value of .10 shows that this β values is not very close to the desired result. Since this is a trial and error process, it is time to select another value of c. The procedure will be repeated until the value that comes as close as possible to the desired β is found. As a guide to this trial and error procedure, the following generalization may be used. *For a given value of α, the larger the c value is, the smaller the value of β will be.*

In subsequent trials, values of c = 4 and c = 5 are tried. For c = 4,

$$nAQL = 2.43$$
$$nLTPD = 7.29$$
$$β = .149 \text{ (too high for the desired β)}$$

For c = 5

$$nAQL = 3.15$$
$$nLTPD = 9.45$$
$$β = .092 \text{ (too low for the desired β)}$$

The values of c = 4 and c = 5 have bracketed the desired value of β. Since it is impossible to have a fractional acceptance number, c must be a whole number. The c value for which β is as close as possible to the desired value of β is selected. For this example, this value is c = 5.

Once the β value has been determined algebra is used to determine the corresponding sample size. Since AQL = .02 and LTPD = .06, either nAQL = 3.15 or nLTPD = 9.45 can be solved for n.

$$nAQL = 3.15$$
$$n(.02) = 3.15$$
$$n = 157$$

or

$$nLTPD = 9.45$$
$$n(.06) = 9.45$$
$$n = 157$$

The sampling plan that comes closest to meeting the risks of α = .10, β = .10 and the desired quality levels of AQL = .02 and LTPD = .06, is one in which a random sample of 157 parts is taken and the entire lot is accepted if there are five or fewer defects in the sample.

When the Poisson distribution is used to develop sampling plans, no mention is made of lot size. This is because the sample is relatively small compared to the lot. When this is the case, the representative sample is used to approximate the lot. If the sample size approaches the lot size, the need for sampling should be carefully examined.

Following is another example of the calculation trial and error procedure involved in developing a sampling plan to fit a given producer's risk, consumer's risk, AQL, and LTPD. Note particularly the special relationship between α and AQL and β and LTPD.

Example 10.6

Develop a single sample sampling plan to meet these conditions:

$$AQL = .01 \quad LTPD = .05 \quad \alpha = .05 \quad \beta = .15$$

Solution

In order to develop a sampling plan, n and c must be identified. The first step is to determine the probability of accepting acceptable material, or $1 - \alpha$. This is .95. The second step is to calculate the ratio of LTPD to AQL. This ratio is 5.

The trial and error procedure cannot begin with the selection of a value for c of c = 1. The c = 1 column in the cumulative Poisson table, Table I, is used to find the $1 - \alpha$ value. The value that comes closest to .95 corresponds to an np' of nAQL value of .35. (Note: Although a more precise value could be found through interpolation, the extra effort generally does not add to the resulting sampling plan.) The LTPD/AQL ratio of 5 gives an nLTPD value of (5)(.35) = 1.75. The corresponding probability of accepting unacceptable material, β, is .478. This is larger than the desired β value of .15.

A larger value of c is tried next. Remember, the idea is to bracket the desired value of β, and then select the c for which β comes closest to the desired value.

For c = 4,

$$nAQL = 1.95$$
$$nLTPD = 9.75$$
$$\beta = .035 \quad \text{too low}$$

For c = 3

$$nAQL = 1.35$$
$$nLTPD = 4.0$$
$$\beta = .097 \quad \text{too low}$$

For c = 2

$$nAQL = .80$$
$$nLTPD = 4.0$$
$$\beta = .238 \quad \text{too high}$$

The desired value of β falls between acceptance numbers of 2 and 3. When c = 3 the β value comes closer to the desired value of .15. The sampling plan is completed by solving for n.

$$nLTPD = 6.75$$
$$n(.05) = 6.75$$
$$n = 135$$

A random sample of 135 and an acceptance number of 3 come closest to meeting the conditions desired.

A brief discussion of the meaning of AQL and LTPD is appropriate at this point. AQL, which is a measure of the percentage of defective material, in effect specifies the desired quality of the material. It is a goal. Organizations would like to accept material that is AQL percent defective or better. LTPD defines the quality of material that is unacceptable and should be rejected. Sampling plans are designed so there will be a small probability of accepting this material. However, sometimes material will be accepted that should not be accepted. For material that is between AQL percent and LTPD percent defective, how high or low the probability of acceptance is depends on the relative steepness of the OC curve. The steeper the curve, the smaller the in-between area corresponding to acceptance of such material.

A sampling plan does not guarantee that only 100 percent perfect material will be accepted. Some defective material will be accepted, although the amount can be minimized by proper specification of the β risk. The advantage of using a sampling plan is that then one usually has a reasonably good idea how much unacceptable material will be accepted.

Double Sample Sampling Plans

Although the single sample sampling plan is the easiest of the acceptance sampling plans to implement, the most frequently used plan is the double sample sampling plan. This plan involves taking two samples. The first sample is used to identify material of either very high or very low quality. Material that is borderline, or of marginal quality, merits another look, so a second sample is used to get a better evaluation of the overall quality.

> A double sampling procedure was chosen as the basic (multiple) standard procedure for two reasons: first, because it is substantially more efficient, requiring on the average a lesser amount of inspection than single in discriminating between good and bad quality; and second, because it had been found to have certain psychological advantages, appearing, for example, to give a doubtful lot a second chance to get by. (Samson et al., 1970, p. 159)

When material is either very good or very bad, a double sample sampling plan is more economical than a comparable single sample sampling plan. Inspection of lots of marginal quality generally takes more time — and therefore costs more — with a single sample plan than with a double sample plan that has approximately the same OC curve. The use of any multiple sample sampling plan has the potential to reduce an organization's inspection costs, and thus should be seriously considered by management.

A double sample plan has the same α, β, AQL, and LTPD as the single sample sampling plan. A double plan has two samples, which are identified as n_1 and n_2, and two acceptance numbers, which are called c_1 and c_2. A double sample sampling plan is used as follows:

In a given lot, N, a first random sample of size n_1 is inspected. If c_1 or fewer defects are found, the lot is accepted because the material is very good. If more than c_2 defects are found in the first sample, the lot is rejected because there is evidently very poor quality material present.

When the number of defects in the first sample is more than c_1 but less than or equal to c_2, a decision is made to not decide. In the case of such questionable material, the procedure is to look at more data or additional inspection results before making a decision. A second sample of size n_2 is taken. If the *total* number of defects found in *both* samples exceeds c_2, then the lot is rejected. Figure 10.6 is a flowchart of the process.

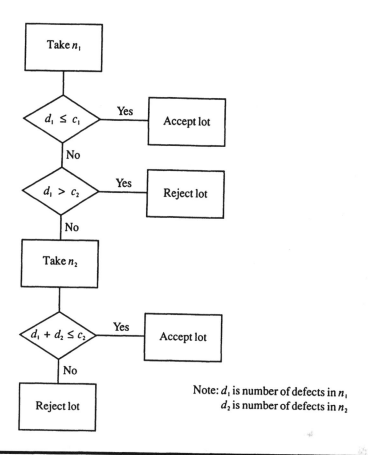

Figure 10.6 Logic for a Double Sample Sampling Plan

Example 10.7

Describe the sampling and decision making procedure in a double sample sampling plan defined as follows:

$$n_1 = 100 \quad n_2 = 75 \quad c_1 = 2 \quad c_2 = 4$$

Solution

The inspector takes the first random sample of 100 items. If, in this first sample, there are zero, one, or two defects, the lot should be accepted. If there are five or more defects, the lot should be rejected. If three of four defects are found in the first sample, the lot may still be accepted if one or zero defects, respectively, are found in the second sample.

As long as the total number of defects found in n_1 and n_2 does not exceed c_2, the material is acceptable. When the cumulative number of defects exceeds c_2, the material is no longer acceptable and must be rejected. Table 10.3 shows various possibilities for the inspection results.

Table 10.3

Defects in n_1	Defects in n_2	Action
0	—	Accept
1	—	Accept
2	—	Accept
3	0	Accept
3	1	Accept
3	2	Reject
4	0	Accept
4	1	Reject
5	—	Reject

As can be seen, lots of either very high or very low quality are disposed of on the first sample. Only the intermediate quality lots require a second look. The double sample sampling plan gives an inspector the opportunity to obtain more data before making a decision.

OC Curves for Double Sample Sampling Plans

A double sample sampling plan, like a single plan, can be represented graphically by an OC curve. The OC curve for a double sample sampling plan shows the probability of accepting a lot with any given percentage of defective material.

Calculating probability values is a bit more complicated for a double sample sampling plan than for a single sample sampling plan. There are two samples to worry about, and the probability of accepting is dependent on the results of the two sample inspections. Some symbols used in developing the OC curve for a double sample sampling plan are provided here:

- Pa_1 is the probability of accepting the lot based on inspection of only the first sample.
- Pa_2 is the probability of having to take a second sample and accepting the lot based on the inspection results from the second sample.
- Pa is the probability of accepting the lot based on the sample inspection. Pa is the sum of Pa_1 and Pa_2.

The development of the OC curve is based on the fundamental rules of probability. If the logic behind these rules is not obvious, Chapter 3 of this text should be reviewed.

Example 10.8

Given the double sample sampling plan in which $n_1 = 25$, $n_2 = 25$, $c_1 = 2$, and $c_2 = 5$, determine the probability of accepting the lot (a) in general terms and (b) when the material is 4 percent defective as determined by a process capability study.

Solution

(a) To generalize the solution it is necessary to write expressions identifying Pa_1 and Pa_2. Pa_1 is defined as the probability of accepting the lot based on inspection of the first sample:

$$Pa_1 = P(2 \text{ or fewer defects in } n_1)$$

Pa_2 is defined as the probability of taking a second sample and then accepting the based on the results of the second sample inspection. In this example, a second sample will be taken if there are exactly three or four or five defects in n_1. If there are exactly three defects in the first sample, then there can be two or fewer defects in n_2. If there are exactly four defects in n_1, then there can be one or fewer defects in the second sample. If there are exactly five defects in the first sample, the lot will be accepted only if zero defects are found in the second sample.

Written in the standard form, the formula is

$$Pa_2 = P(\text{exactly 3 defects in } n_1)P(2 \text{ or fewer defects in } n_2)$$
$$+ P(\text{exactly 4 defects in } n_1)P(1 \text{ or fewer defects in } n_2)$$
$$+ P(\text{exactly 5 defects in } n_1)P(0 \text{ defects in } n_2)$$

Once Pa_1 and Pa_2 have been identified, the probability of accepting the lot is calculated by adding the two component parts:

$$Pa = Pa_1 + Pa_2$$

(b) Calculation of the probability of accepting a lot using the double sample sampling plan if the material is actually 4 percent defective involves the use of the Poisson probability distribution. It will require the use of both the cumulative distribution, found in Table I, and the exact distribution, found in Table B.

The following rules govern the use of these tables as relates to probabilities of acceptance:
- The exact Poisson table, Table B, is used whenever the expression *exactly* appears in the probability statement.
- The cumulative Poisson table, Table I, is used when the expression *or fewer* or the expression *or less* appears.

Two additional rules specify use within the appropriate table:
- n_1p' is required to identify probabilities that reference the first sample.
- n_2p' is required to identify probabilities that reference the second sample.

The calculations for this example begin with the determination of the np' values.

$$n_1p' = (25)(.04) = 1.0$$
$$n_2p' = (25)(.04) = 1.0$$

The probabilities can now be calculated for the expressions developed earlier for Pa_1 and Pa_2.

$Pa_1 = P(2 \text{ or fewer defects in } n_1) = .920$ from table I
with $n_1p' = 1.00$ and $c = 2$

$Pa_2 = P(\text{exactly 3 defects in } n_1)P(2 \text{ or fewer defects in } n_2)$
$+ P(\text{exactly 4 defects in } n_1)P(1 \text{ or fewer defects in } n_2)$
$+ P(\text{exactly 5 defects in } n_1)P(0 \text{ defects in } n_2)$

$$Pa_2 = (.0613)(.920) + (.0153)(.736) + (.0031)(.368)$$
$$= .056 + .011 + .001$$

$$Pa_2 = .068$$

$$Pa = .920 + .068 = .988$$

The probability of accepting a lot of material that is 4 percent defective is .988 when this sampling plan is used. The value .988 would be one point on the plan's operating characteristic curve.

Example 10.9

Plot the entire OC curve for the sampling plan in example 10.8, in which $n_1 = 25$, $n_2 = 25$, $c_1 = 2$, and $c_2 = 5$.

Solution

Several points are required to graph the OC curve, so various values of p' must be selected and the corresponding values of Pa determined. The two obvious values of Pa = 1.0 when $p' = 0$ and Pa = 0 when $p' = 1.0$ can immediately be utilized. Another point that can be used is the value of Pa for $p' = .04$, which was calculated to be .988 in Example 10.8. For $p' = .06$ the value of Pa is calculated as follows:

$$Pa_1 = .809$$

$$Pa_2 = (.1255)(.809) + (.0471)(.588) + (.041)(.223)$$
$$= .102 + .026 + .003 = .131$$

$$Pa = .809 + .131 = .940$$

Table 10.4 tabulates values of Pa for other values of p'.

After these values have been calculated, the OC curve shown in Figure 10.7 is graphed. The shape of this OC curve is rather broad, indicating that this plan favors the producer and it might be advantageous for the consumer to try another double sample sampling plan.

A logical extension of the double sample sampling plan is the multiple sample sampling plan. The multiple sample sampling plan at first seemed to

Table 10.4

p'	Pa_1	Pa_2	Pa
0	1	—	1
0.02	0.986	0.013	0.999
0.04	0.920	0.068	0.988
0.06	0.809	0.131	0.940
0.08	0.677	0.164	0.841
0.16	0.238	0.070	0.308
1	0	—	0

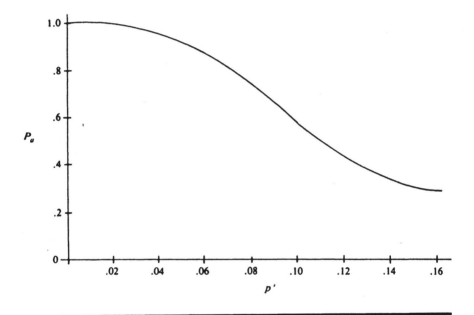

Figure 10.7   OC Curve for Example 10.9

extend the advantages of double sample sampling, particularly in terms of the cost factors.

> However, a few early trials brought out factors that had not been foreseen, factors relating to the human element. Inspectors did not like the added record keeping for one thing, but principally, they reacted unfavorably to multiple sampling for another reason — they were opposed, as some said, "to a plan that can't make up its mind." (Dodge, 1969, p. 87)

There are some multiple sample sampling plans in use, but according to published reports, they account for less than 5 percent of all sampling applications (Dodge, 1969, p. 87).

Comparison of Sampling Plans

Comparison of sampling plans usually focuses on two factors — overall quality after inspection of many lots of material and inspection costs. The following sections will examine these important concepts.

Rectifying Inspection

Up to this point, it has been assumed that when material does not meet quality standards, it is rejected. This is true if rejection is defined as a setting aside of the material. However, it should not be assumed that material that does not meet specifications is immediately scrapped when a sampling plan is used. Very often an organization can salvage material that does not initially meet the desired quality standards by subjecting the material to rectifying inspection.

> **Rectifying inspection** is the process of conducting a 100 percent reinspection of a rejected lot and replacing the defective or nonconforming parts with "acceptable" products.

The process will improve the quality of the material, but it will not make the material perfect, given the flaws of 100 percent inspection discussed earlier in this chapter. The quality of the material after rectifying inspection will depend on the quality of the material before rectifying inspection. If the quality beforehand is known, either from experience or from the results of a process capability study, the quality afterward can be calculated in a straightforward manner.

> The **average outgoing quality,** or **AOQ,** of a lot is the expected, or average, quality of the material after rectifying inspection has been performed. (Note: Remember that this is an average value. Individual values for specific lots may vary from this descriptive statistic.)

Each time a lot undergoes rectifying inspection the AOQ improves, but by a smaller amount each time. The average outgoing quality when rectifying inspection is performed is calculated using the following formula:

$$AOQ = (Pa)(p')$$

The formula is the same for single sample sampling plans and double sample sampling plans. The value of AOQ is useful only if the value of p' is known ahead of time, for both AOQ and p' are needed to calculate Pa.

In many cases the initial quality of a lot of material rejected by a sampling plan is not known, yet it is desirable to know what the quality will be after a rectifying inspection is performed.

The **average outgoing quality limit,** or **AOQL,** is the maximum expected value of AOQ for any sampling plan. The AOQL is the worst quality, in terms of percentage defective, that can be expected in a lot of material that has undergone rectifying inspection, regardless of the quality of material before rectifying inspection. (Note: Remember that this is an average value. Individual values for specific lots may vary from this descriptive statistic.)

The easiest way to determine the AOQL is by graphing the average outgoing quality as a function of the percent defective. The AOQL is the maximum point on the curve.

Example 10.10

Calculate the values of AOQ corresponding to initial percentages defective of 0. 2. 4. 6. 8. 10, 11, and 12 percent and determine the AOQL for the sampling plan with the following parameters:

$$n_1 = 25 \quad n_2 = 25 \quad c_1 = 2 \quad c_2 = 5$$

Solution

The OC curve for this plan was developed in Example 10.9. The easiest way to determine the values of AOQ and AOQL is by making a table of the already determined values of Pa and the respective values of p'. An additional column showing AOQ is added to the table. Table 10.5 shows these values.

The largest value of AOQ will occur if the lot is 10 percent defective. Thus if a lot were inspected with the above double sample sampling plan and subjected to rectifying inspection, it would never be worse than 7.05 percent defective regardless of the initial quality.

Table 10.5

p′	Pa	AOQ
0.00	1.000	0.0000
0.02	0.999	0.0199
0.04	0.988	0.0395
0.06	0.940	0.0564
0.08	0.841	0.0672
0.09	0.774	0.0696
0.10	0.705	0.0705*
0.11	0.632	0.0695
0.12	0.556	0.0667

*AOQL

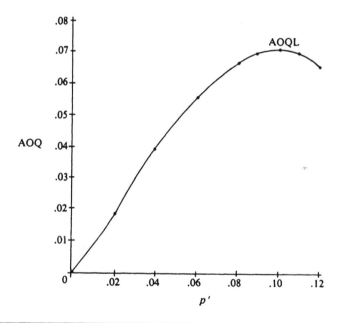

Figure 10.8 AOQ Curve Showing AOQL for Example 10.10

Example 10.11

Calculate AOQL for the single sample sampling plan with the following parameters:

$$n = 120 \quad c = 4$$

Solution

The tabular approach is again the most straightforward. Table 10.6 shows the values of p′, Pa, and AOQ. The AOQL, indicated with an asterisk, is 2.12 percent.

Table 10.6

p′	Pa	AOQ
0.02	0.904	0.018
0.04	0.476	0.019
0.05	0.285	0.01425
0.03	0.706	0.212*

*AOQL

Any material, regardless of its original quality, will be no worse than 2.12 percent defective, on average, after undergoing rectifying inspection, if necessary, under this single sample sampling plan.

If after undergoing rectifying inspection, the AOQL meets the desired quality specifications, all is well. However, sometimes more than one rectifying inspection has to be done on a critical product if the first does not meet the goal. Each successive inspection increases the quality, but by less than the preceding inspection. When inspection is used in this fashion, its purpose is to remove as many poor quality products as possible.

Example 10.12

In example 10.11, where n was 120 and c was 4, the value of AOQL was .0212 and it occurred when p′ was 3 percent. After rectifying inspection, the percentage defective would be 2.12 percent. What would be the effect of a double and triple rectifying inspection?

Solution

The probability of accepting a 2.12 percent defective lot using the sampling plan is .884. The new value of AOQ and hence AOQL, is

$$AOQ = (.884)(.0212) = .0187$$

This means the material will be, at its worst, 1.87 percent defective after having gone through a double rectifying inspection. If the desired quality conditions still have not been met, another rectifying inspection may be performed on material that is than no worse than 1.87 percent defective. If this is done, the new value of Pa corresponding to a p' of .0187 is .922. The new value of AOQ, and hence AOQL, is

$$AOQ = (.922)(.0187) = .0172$$

The material will be, on average, at its worst, 1.72 percent defective after having gone through a triple rectifying inspection.

Rectifying inspection can be used, on rare occasions, to find all the defective parts in a lot. AOQ and AOQL will measure how successful this effort is. However, *inspecting quality into a product is not a good practice.* It is much better to make the product right in the first place. The goal should be to build quality into the product rather than to inspect quality into the product.

Inspection Costs

Although the AOQL gives an indication of the quality after the rectifying inspection, it is not necessarily indicative of the costs involved in performing the inspection. There are two statistics that can be calculated that will give an estimate of the potential inspection costs for any sampling plan — the average total inspection and the average sample number.

Over the long run the average number of parts that might be inspected may be significantly different for two similar acceptance sampling plans. Costs of inspectors' time, testing equipment usage, and even material itself — if the test required by the inspection is destructive — can mount up. Naturally it is advisable to keep costs as low as possible without sacrificing the ability to meet the quality standards. Additionally, any reduction in inspection costs represents a possible source of resources for process improvements.

For comparing sampling plans that have essentially the same risks, there are two probability-based parameters that give estimates of the average number of items that will be inspected under certain circumstances.

The **average total inspection,** or **ATI,** is the average number of parts, for a specific value of p', that is inspected when a rectifying inspection is performed.

The ATI is used when rectifying inspection is being performed to ensure that the quality of material passing into or out of the system meets certain standards. The ATI depends on whether or not each lot is accepted by a sampling plan and on the lot size itself. If a lot with an estimated value of p' is accepted, then the ATI for that lot will be the sample size. If the lot is rejected the entire lot will be reinspected and the ATI will be the lot size. Whether the ATI over time is closer to the sample size or the lot size will depend on the original percentage defective. For a single sample sampling plan the average total inspection is calculated with the following formula:

$$ATI = nPa + N(1 - Pa)$$

For a double sample sampling plan the ATI is represented by the following relationship:

$$ATI = n_1Pa_1 + (n_1 + n_2)Pa_2 + N(1 - Pa)$$

Example 10.13

Calculate the ATI for the single sample sampling plan with the following parameters:

$$n = 100 \quad c = 3 \quad N = 800$$

Solution

Each value of p' will have a different value of ATI, since the ATI depends on the probability of performing a rectifying inspection. The probability of accepting a lot that is 3 percent defective with this sampling plan is .647. The ATI is calculated as follows:

$$ATI = (100)(.647) + 800(1 - .647) = 347.1$$

If the material is 3 percent defective, an average of just over 347 parts will be inspected when rectifying inspection is performed. Additional ATIs for the sampling plan are tabulated in Table 10.7. As can be seen, the worse the quality of the material, the greater the ATI. This is logical; the higher the percentage defective, the more likely it is that the material will be rejected and subjected to a rectifying inspection.

Table 10.7

p′	ATI
0.00	100
0.03	347
0.06	694
0.09	785
0.15	800
1.00	800

Example 10.14

Calculate the ATI for the sampling plan with the following parameters. Use a p′ of 5 percent.

$$n_1 = 75 \quad n_2 = 25 \quad c_1 = 2 \quad c_2 = 4 \quad N = 5{,}000$$

Solution

The probability of accepting a 5 percent defective lot with this sampling plan is determined as follows:

$$Pa_1 = P(2 \text{ or fewer defects in } n_1) = .2775$$

$$Pa_2 = P(\text{exactly 3 defects in } n_1)P(1 \text{ or fewer defects in } n_2)$$
$$+ P(\text{exactly 4 defects in } n_1)P(0 \text{ defects in } n_2)$$

$$Pa_2 = (.2066)(.645) + (.1937)(.287) = .1888$$

$$Pa = Pa_1 + Pa_2 = .2775 + .1888 = .4663$$

The ATI is then calculated using these values:

$$ATI = 75(.2775) + (75 + 25)(.1888) + 5000(1 - .4663) = 2708$$

An average of 2708 parts must be inspected to ensure that the material is of acceptable quality.

The **average sample number,** or **ASN,** is the average number of pieces, for a specific value of p′, that is inspected when a rectifying inspection is not performed. It is the average number of pieces that must be inspected before a decision is made to accept or reject.

Rarely will the ASN be calculated before the acceptability of the lot is decided. The value of the ASN depends on whether the full sample is inspected or less than the full sample is evaluated. It is not, as the old saying goes, necessary to try each apple to know that the entire barrel is rotten.

For a single sample sampling plan the average sample number is defined by the equation:

$$ASN = n$$

For a double sample sampling plan, the average sample number is defined by the following:

$$ASN = n_1 + P(taking\ n_2)(n_2)$$

Example 10.15

Calculate the average sample number for the double sample sampling plan described as follows:

$$n_1 = 50 \quad n_2 = 50 \quad c_1 = 1 \quad c_2 = 3$$

Solution

Each value of p′ will have a different ASN. The ASN will be calculated for several values of p′ and the results will be graphed to show how ASN acts as a function of p′.

To calculate the ASN it is necessary to know the probability of taking a second sample. A second sample is necessary when the lot is neither accepted nor rejected on the basis of the first sample inspection. For this sample, the lot will be accepted if the first sample has one or fewer defects. This means a second sample will be taken if there are exactly two or exactly three defects in the first sample. The formula for ASN can be rewritten to fit this example.

$$ASN = n_1 + P(\text{exactly 2 or exactly 3 defects in } n_1)(n_2)$$

If p' is equal to 2 percent, the probabilities, as taken from the exact Poisson table (Table B).

P(exactly 2 defects in n_1) = .1839
P(exactly 3 defects in n_1) = .0163
P(exactly 2 or 3 defects in n_1) = .1839 + .0613 = .2452

Thus the ASN is calculated for $p' = .02$:

$$ASN = 50 + (.2452)(50) = 62.26$$

An average of just slightly more than 62 parts must be inspected with this sampling plan before a decision can be made as to whether to accept or reject the lot based on the sample inspection. Sometimes it may be necessary to inspect only 4 parts out of the first sample before rejecting the lot. Other times all 100 pieces in both samples will be inspected before the lot is judged acceptable. But over the long run, 62.26 parts is the best estimate of how many pieces will be inspected before a decision is reached, given that the material is 2 percent defective to start with. This value can be used to calculate the average expected inspection cost.

Table 10.8 tabulates values of ASN for other values of p'. They are than graphed in Figure 10.9 to show an overall picture of the ASN.

Table 10.8

p'	ASN
0.00	50
0.02	62
0.04	73
0.08	67
0.16	52
0.25	50

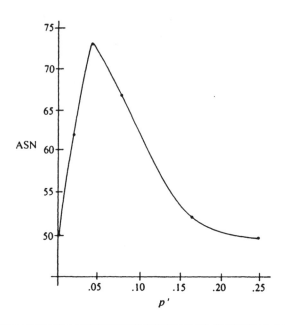

Figure 10.9   ASN Curve for Example 10.15

The larger the value of ASN, the greater the cost of inspection. If more pieces, on average, must be inspected, more inspection costs will be incurred. When sampling plans are considered, the plan with the lower ASN is usually chosen. Economy cannot be neglected. The ASN and the ATI allow the comparison of sampling plans on a quantitative or financial basis. This is, of course, important to creating a quality product.

Published Sampling Plans

Not all sampling plans are developed by a particular company's quality control department. This section describes some of the most frequently used published sampling plans. Special attention is given to the plan which used to be known as Mil-Std 105, but is now ANSI Z1.4-1993. It is published by the American Society for Quality Control (ANSI/ASQC Z 1.9-1993).

ANSI/ASQC Z 1.4-1993

This standard is a widely used and accepted attribute sampling plan. The key quality indicator is the AQL. The plan offers:

- A choice of 26 AQL values
- The probability of accepting at AQL quality varies from 89 to 99.5 percent
- Defects are classified as critical, major, or minor.

> The purchaser also specifies the relative amount of inspection or inspection level to be used. For general applications there are three levels, and the level II is regarded as normal. Level II is generally used unless factors such as the simplicity and cost of the item, inspection cost, destructiveness of inspection, quality, consistency between lots, or other factors make it appropriate to use another level. (Juran, 1988, p. 25.44)

The first step in use of the standard is determination of the lot size, which is generally a function of production. The sample size depends on the specified AQL. The type of sampling, single, double, or multiple, must also be specified, as must the inspection level. When this information is known a code letter is obtained from the master table within the sample. Knowing the code letter, the AQL, and the type of sample, the details of the sampling plan, that is the n and the c, are read.

The standard also includes the following descriptive information for each sampling plan:

- ASN curves
- OC curve and Tabulated Values
- AOQ values (by default)

Also included in the standard is a system for monitoring the ongoing quality of incoming material. Switching rules indicate when reduced or tightened inspection should be considered.

ANSI/ASQC Z1.9-1993

This sampling standard is titled *Sampling Procedures and Tables for Inspection by Variables for Percent Nonconforming*. The standard was prepared to meet a growing need for the use of the standard sampling plans for inspection by variables.

> The variables sampling plans apply to a single equality characteristic which can be measured on a continuous scale, and for which quality is expressed in terms of percent nonconforming. In comparison with

attributes sampling plans, variables plans have the advantage of usually resulting in considerable savings in sample size for comparable assurance as to the correctness of decisions in judging a single quality characteristic. (ANSI/ASQC Z1.9-1993)

The standard includes the following:

- Section A provides general procedures for the standard
- Sections B and C describe specific procedures and applications when variability is unknown
- Section D describes plans when variability is known
- Operating characteristic curves

Attribute plans have the advantage of greater simplicity, of being applicable to either single or multiple quality characteristics, and of requiring no knowledge about the distribution of the continuous measurements of any of the quality characteristics.

Dodge-Romig Tables

The Dodge-Romig tables for attribute-based sampling plans were developed originally for Bell Labs and Western Electric. They are designed to minimize the ATI; however, when lots are rejected and a rectifying inspection must be performed, their benefit is minimized. The plans are identified by AOQL and by LTPD. There are no multiple sample sampling plans provided.

Four different sets of tables are provided:

- Single sample lot tolerance tables
- Double sample lot tolerance tables
- Single sample AOQL tables
- Double sample AOQL tables

The plans are based on a consumer's risk of 10 percent. They tend to minimize inspections costs, but do tend to offer a very limited set of inspection plans.

When they are used, the first step is determination of the lot size. Next the desired LTPD is specified to accompany the 10 percent β risk. The plan is then identified and used.

Summary

This chapter has just skimmed the surface of acceptance sampling, illustrating the power that acceptance sampling can have. The chapter has concentrated on the basic statistical procedures rather than the inspection procedures and quality assurance program that must accompany the inspection plans.

The reader should now have a reasonable understanding of the fundamental statistical quality control functions that go hand in hand with the idea of acceptance sampling. However, it is important to keep in mind that these statistical procedures only provide guidance to the person who must actually make the final decision. Many other factors, such as cost and time, must also be considered. Never should decisions be made automatically or blindly just because some number indicates that a particular course is best.

References

ANSI/ASQC Z1.4-1993, American Society for Quality Control, Milwaukee, 1993.
ANSI/ASQC Z1.9-1993, American Society for Quality Control, Milwaukee, 1993.
Dodge, H.F., "Notes on the Evolution of Acceptance Sampling Plans, *Journal of Quality Technology,* April, 1969
Juran, J. (Ed.), *Quality Control Handbook,* 4th ed., McGraw-Hill, New York, 1988.
Samson, C., Hart, P., and Rubin, C., *Fundamentals of Statistical Quality Control,* Addison Wesley, Reading, MA, 1970.

Practice Problems

1. A manufacturer uses a sampling plan in which n = 145 and c = 3. What is the probability of accepting a lot with this plan if $p' = 4$ percent?
2. What is the probability of accepting a lot that is actually 2 percent defective with a sampling plan in which n = 250 and c = 5?
3. What is the probability of accepting a lot that is actually 5 percent defective with a sampling plan in which n = 100 and c = 3?
4. What is the probability of accepting a lot that is actually 2 percent defective with a sampling plan in which n = 400 and c = 6?
   In problems 5, 6, and 7 develop a single sample sampling plan that comes closest to meeting the following conditions.
5. $\alpha = .05$ $\beta = .05$ AQL = .025 LTPD = .075

6. $\alpha = .05$ $\beta = .20$ AQL = .025 LTPD = .075
7. $\alpha = .05$ $\beta = .10$ AQL = .030 LTPD = .090
8. A manufacturer wishes to accept material that is 1 percent defective. He would like to be assured of this 95 percent of the time. On the other hand, his engineer has assured him that 4 percent defective material is acceptable 12 percent of the time. Develop a single sample sampling that comes closest to meeting these conditions.
9. A manufacturer has entered into a contract with her customer. The quality terms state that the producer's risk is 8 percent, the consumer's risk is 12 percent, the acceptable quality level is .5 percent, and the lot tolerance percent defective is 3 percent. What single sample sampling plan will come closest to meeting these terms?
10. Develop a single sample sampling plan for the following conditions.

$$\alpha = .05 \quad \beta = .05 \quad AQL = .02 \quad LTPD = .10$$

What is the chance of accepting a lot that is 4 percent defective with the sampling plans shown in problems 11 and 12?

11. $n_1 = 100$ $n_2 = 100$ $c_1 = 4$ $c_2 = 6$
12. $n_1 = 100$ $n_2 = 100$ $c_1 = 3$ $c_2 = 5$
13. Calculate the AOQL for each of the following single sample sampling plans.
    (a) $n = 150$ $c = 5$
    (b) $n = 400$ $c = 6$
    (c) $n = 6$ $c = 1$
    (d) $n = 400$ $c = 4$
14. Calculate the AOQL for the following double sample sampling plans:
    (a) $n_1 = 200$ $n_2 = 100$ $c_1 = 2$ $c_2 = 3$
    (b) $n_1 = 70$ $n_2 = 30$ $c_1 = 2$ $c_2 = 4$
15. Calculate the ASN for the following single sample sampling plans:
    (a) $n = 600$ $c = 10$ $p' = .02$
    (b) $n = 100$ $c = 6$ $p' = .10$
16. Calculate the ASN for the following double sample sampling plans:
    (a) $n_1 = 67$ $n_2 = 33$ $c_1 = 3$ $c_2 = 5$ $p' = .05$
    (b) $n_1 = 48$ $n_2 = 48$ $c_1 = 2$ $c_2 = 5$ $p' = .05$
    (c) $n_1 = 100$ $n_2 = 100$ $c_1 = 2$ $c_2 = 6$ $p' = .03$
    (d) $n_1 = 150$ $n_2 = 100$ $c_1 = 3$ $c_2 = 7$ $p' = .05$
    (e) $n_1 = 70$ $n_2 = 80$ $c_1 = 3$ $c_2 = 6$ $p' = .06$

17. Calculate the ATI for the following single sample sampling plans:
    (a) $n = 105$ $c = 3$ $N = 1000$ $p' = .04$
    (b) $n = 70$ $c = 3$ $N = 1000$ $p' = .05$
    (c) $n = 100$ $c = 3$ $N = 3000$ $p' = .06$

18. Calculate the ATI for the following double sample sampling plans:
    (a) $n_1 = 100$ $n_2 = 50$ $c_1 = 2$ $c_2 = 3$ $N = 1000$ $p' = .04$
    (b) $n_1 = 50$ $n_2 = 50$ $c_1 = 2$ $c_2 = 4$ $N = 3000$ $p' = .06$
    (c) $n_1 = 100$ $n_2 = 100$ $c_1 = 2$ $c_2 = 6$ $N = 2400$ $p' = .03$

19. A rule of thumb double sample plan is described as follows: Select a random sample of size two from a lot. If both pieces are good, accept the lot. If both pieces are bad, reject the lot. If only one piece is good, take a second sample of one piece. If that piece is good, accept the lot. If that piece is bad, reject the lot. What is the probability of accepting a lot with this sampling plan if the material is 25 percent defective?

20. A rule of thumb sampling plan tells the inspector to take a first sample of size 100. If there are two or fewer defects in the sample, accept the lot. If there are between three and eight defects in the first sample, take a second sample also of size 100. If the number of defects in the first and second samples combined is less than seven, the lot should be accepted. If the combined number of defects is nine or more, the lot should be rejected. If there are seven or eight defects in the first two samples, a third sample of 50 should be taken. If the total number of defects found in all three samples is nine or greater, the lot should be rejected. What is the probability of accepting a lot with this sampling plan if the material is actually 4 percent defective?

21. A manufacturer has just entered into a contract with a supplier. The manufacturer would like to accept material that is no worse than 2 percent defective. He will determine whether to accept or reject material by using a sampling plan that calls for him to inspect 100 parts in each lot and accept if there are four or fewer defects in the sample. This plan will ensure that bad material is accepted only 10 percent of the time and that good material is rejected only 5 percent of the time. What is the highest percentage of defective material that the supplier can offer and still hope to meet the terms of the contract?

11 Reliability

Introduction

Although producers and consumers of products have always been concerned about the reliability of the goods produced and consumed, only recently has there been a significant interest in quantifying the study of reliability. Any product's reliability is a result of how the product was produced. The reliability depends on the quality of every step, from the appropriateness of the design to the consistency of the production process.

Nothing lasts forever. Not even Oliver Wendell Homes' famous carriage, the "one hoss shay,"

> ...That was built in such a logical way
> It ran a hundred years to a day.

Not all products last a hundred years. Some products seem to wear-out before they even lost that brand new smell or feel. Just how long a product will last and perform the task it was designed to do is indicative of the reliability of the product. A formal definition for reliability follows:

Reliability is the probability that a product will perform the intended function, in the manner specified, for a particular period of time.

This definition, although straightforward, has some significant implications. First is the idea that the function for which a product is intended must be defined. Reliability is dependent on the use of the product. An automobile used by a little old lady only on Sundays will have a different reliability than the same car used as a taxi or a police car. A television that was originally designed for use by a mature couple in their living room but ends up being used as the host for a video game system may have been just fine for the original use but may not be able to handle the different kinds of stresses placed on it by the alternative uses.

Another factor affecting the reliability of any product is the maintenance given to it. Commercial airliners are constantly undergoing preventive maintenance. Components are refurbished or replaced before they break down. Thus air travel is quite safe and reliable. However, other products, such as the family washing machine, are often maintained only after they break down. Preventive maintenance is often dictated by the economics of the situation. The cost of an airplane failure is tremendous, so the benefits of performing regular maintenance outweigh the associated costs and inconveniences. In the home, though, the benefits received from higher reliability are often not viewed as being worth the cost and bother of performing maintenance.

The study of reliability — both the reliability of the individual components and the resulting reliability of the interaction of individual components in what might best be called a system — can and should guide the design, manufacture, and operation of all products. The quality standards established for a product are the result of tradeoffs among considerations of design and use, cost, safety and various other factors.

Measuring Reliability

All products fail sooner or later. Product failure is placed in one of three major categories according to when in the product life cycle it occurs. A failure arising very early in a product's life is called a *wear-in,* or *burn-in* failure. It may be caused by manufacturing problems, defective workmanship, faulty materials, poor design, or a variety of other reasons. The phrase, "If it lasts the first few weeks there won't be any problems at all" is a reference to wear-in failure. To compensate for wear-in failure, sometimes called infant mortality, manufacturers generally provide warranties that cover almost everything during the very early portions of a product's life. The graph in Figure 11.1 shows wear-in failure as a function of time.

The failure of a product that finally wears out and dies from old age is called *wear-out* or *burn-out* failure. No matter how well it is made, and no matter how much maintenance is performed, every product fails eventually. Wear-out failure is often marked by a series of breakdowns that occur after a relatively trouble free life. "Everything was fine until one day the whole thing seemed to fall apart." The phenomenon was perhaps best described by Oliver Wendell Holmes in reference to the "one hoss shay:"

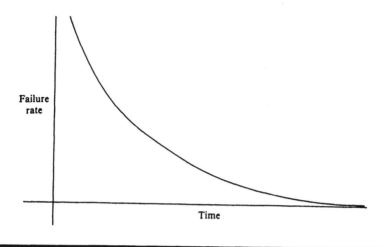

Figure 11.1   Burn-In Failure as a Function of Time

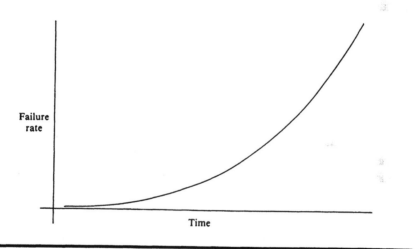

Figure 11.2   Burn-Out Failure as a Function of Time

> You see, of course, if you're not a dunce,
> How it went to pieces all at once,
> All at once, and nothing first,
> Just as bubbles do when they burst.

Figure 11.2 shows a graph of failure rate versus time for burn-out failure. Note the rapid increase in failure rate once the wear-out or burn-out phase begins.

Between the wear-in and wear-out failure is a period of time during which there is normally very little failure, or at least a constant rate of failure. What failure there is, is randomly distributed. This period of relatively failure free operation is called the *steady state operational period.* Figure 11.3 shows the steady state failure rate as a function of time.

When the three graphs showing the three separate failure rate periods are combined, as in Figure 11.4, the picture of the classic reliability function emerges. Because of its shape, this curve is called the "bathtub function."

Reliability experts have empirically determined that the bathtub function can be mathematically represented by the exponential probability distribution. A component's reliability can be calculated once the parameters of the distribution have been identified.

The failure rate, λ, is the average number of products that fail in some specified period of time.

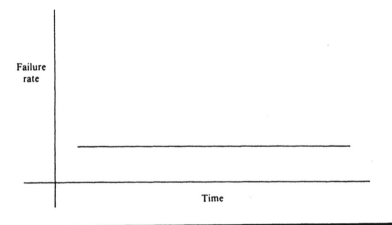

Figure 11.3   Steady-State Failure as a Function of Time

Example 11.1

Aft-Tech wants to purchase a new light bulb that the Light Up Light Bulb Company is promoting. Before deciding on the purchase, though, Aft-Tech wants to know how reliable the bulbs are. In order to determine reliability, they first must calculate the failure rate of the bulbs. In order to determine failure rate, life tests are performed on 10 sample bulbs: Aft-Tech's reliability technician turns on each of the bulbs and records the amount of time that

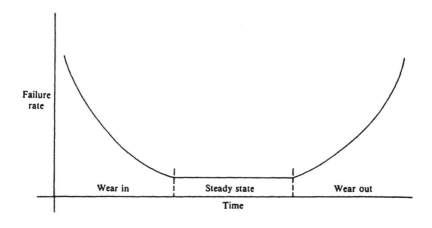

Figure 11.4   The Bathtub Function — Failure Rate as a Function of Time

Table 11.1

Bulb	Hours to Failure
1	400
2	350
3	450
4	400
5	400
6	600
7	300
8	300
9	300
10	500
	4000

elapses before the bulb fails. The results of the life test on the 10 sample bulbs are tabulated in Table 11.1 What is the failure rate?

Solution

The total time required for all ten bulbs to fail was 4,000 hours. On the average, a bulb failed every 400 hours. The number of bulbs that failed each hour, the failure rate, is calculated as follows:

$$\lambda = (10 \text{ bulbs})/(4000 \text{ hours}) = .025 \text{ bulbs/hour}$$

Another parameter need to calculated component reliability is the mean product life. This mean life can be expressed in terms of the mean time to failure or the mean time between failures.

The **mean time to failure, MTTF,** is the average length of time a product performs without failing when maintenance is *not* performed.

The **mean time between failures, MTBF,** is the average length of time a product performs without failing when regular maintenance is performed.

The symbol for both means is μ. The following examples illustrate the difference between the two measures.

Example 11.2

In Example 11.1, Aft-Tech discovered that the ten bulbs lasted a total of 4,000 hours. Determine the mean product life.

Solution

The mean product life would be stated in terms of the MTTF because no maintenance is done on light bulbs. The mean is calculated as follows:

$$\mu = (4000 \text{ hours})/(10 \text{ bulbs}) = 400 \text{ hours/bulb}$$

Example 11.3

Not long ago a certain automobile manufacturer, in order to demonstrate the reliability of its product, boasted of the long life its cars had when only regular routine maintenance was performed — oil changes every 3,000 miles and tuneups every 20,000 miles. Table 11.2 gives the number of miles driven before breakdown for four cars that received this normal servicing. Determine the mean product life and the failure rate.

Solution

The mean time between failures is the average number of miles driven before the auto broke down.

Table 11.2

Car	Miles to Failure
1	120,000
2	135,000
3	110,000
4	125,000
	490,000

$$\mu = (490{,}000 \text{ miles})/(4 \text{ cars}) = 122{,}500 \text{ miles/car}$$

The average failure rate is the average number of cars that fail per mile.

$$\lambda = (4 \text{ cars})/(490{,}000 \text{ miles}) = .0000081 \text{ cars/mile}$$

It should be noted at this juncture that the mean and the failure rate are reciprocals of each other.

Example 11.4

Calculate λ and μ for the data in Table 11.3.

Table 11.3

Number of Units	Hours to Failue	Total Hours
2	4,000	8,000
6	4,500	27,000
8	5,000	40,000
4	5,500	22,000
20		97,000

Solution

The raw data are presented as a frequency distribution. Some of the data have been "grouped" rather than listed individually. The total hours in each case are determined by multiplying the frequency times the hours to failure. The mean is calculated as follows:

$$\mu = (97{,}000 \text{ hours})/(20 \text{ units}) = 4{,}850 \text{ hours/unit}$$

The failure rate is the reciprocal of the mean:

$$\lambda = 1/4{,}850 = .000206 \text{ units/hour}$$

In order to mathematically calculate the reliability of any product it is also necessary to know t, the time period for which failure-free operation is desired. Once all of the terms have been specified, it is possible to mathematically define the reliability of a particular component.

The specific equation to be used depends on the nature of the distribution observed during the life tests. The most common of the distributions are the exponential, the normal, and the Weibull. While exact mathematical tests can be used to measure the goodness of fit, if the shape of the observed distribution, as represented by the frequency histogram, has the shape of the distribution shown in Figure 11.5a, then the exponential distribution is used; if the shape resembles the curve in Figure 11.5b then the normal distribution is used; and if it looks like Figure 11.5c then the Weibull distribution is used. Table 11.4 summarizes the reliability equations for each of these distributions. Other distributions, such as the gamma distribution, occasionally are seen, but so infrequently as to be out of the scope of this text.

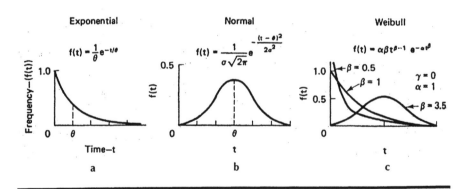

Figure 11.5

Due to ease of calculation we will assume that the exponential distribution is the underlying distribution for all of our examples. In reality, most individual components do indeed follow the exponential distribution. The equation for this is,

Table 11.4

Distribution	Reliability Equation
Exponential	$R = e^{-t/\mu}$
Normal	$R = 1 - \int_0^t f(t)dt$
Weibull	$R = e^{-\alpha\beta}$

$$R = e^{-\lambda t} = e^{-t/\mu}$$

where R is the component reliability and e is the base of the natural logarithms, 2.718. The exponential function appears as Table D.

Example 11.5

A radar scope that Aft-Tech hopes to sell to the FAA for use in commercial airports has a failure rate of .00065 per hour. What is the reliability of this scope for 24 hours of service?

Solution

The calculation for this example uses $\lambda = .00065$ and $t = 24$.

$$R = e^{-(24)(.00065)} = e^{-.0156} = .9845$$

The probability that this scope will operate as intended for 24 hours is .9845.

Example 11.6

Aft-Tech's automobile manufacturing subsidiary has developed a new fuel efficient automobile engine that has a mean time between failures of 15,000 miles. What is the reliability of this engine for (a) 3000 miles and (b) 6000 miles?

Solution

$$\text{(a) } R = e^{-(3000/15,000)} = .8187$$

$$\text{(b) } R = e^{-(6000/15,000)} = .6703$$

Example 11.7

A transistor produced by the Aft-Tech transistor plant is known to have a reliability of .9995 for 4000 hours of operation. What is the failure rate for this transistor?

Solution

In order to determine the value of λ, the reliability function must be used. This time the failure rate rather than the component reliability is the unknown:

$$.9995 = e^{-(4000)(\lambda)}$$

To solve for λ, the natural log (ln) of both sides of the equation must be taken.

$$\ln(.9995) = (-4000)(\lambda)\ln(e)$$
$$-.0005 = (-4000)(\lambda)(1)$$
$$(-.0005)/(4000) = \lambda = .000000125/\text{hour}$$

Series Reliability

Most products do not consist of just one component; they are assemblies of numerous components. The reliability of any product is a function of the interrelationship of the reliabilities of the individual components.

One common way for an assembly to be constructed is with the components in a series relationship. *The reliability of a system of components arranged in a series configuration is dependent on the performance of each component in this configuration. If any of the components in a series configuration fails, then the entire system will fail.*

The series reliability concept is perhaps best illustrated by those strings of Christmas tree lights wired so that if one light fails, the entire string fails to light — successful operation of this system is dependent on performance of all of the components.

In probability terms, for n components arranged in series to reliably function, component 1 *and* component 2 *and* component 3 *and* component n must function. According to the laws of probability, the probability that the entire system will function is the product of the reliabilities of the individual components. If the reliability of the system is designated as R_s, then for n components in a series, the reliability is

$$R_s = R_1 R_2 R_3 ... R_n$$

Example 11.8

Components with the following reliabilities are arranged in series configuration:

$$R_s = .99 \quad R_2 = .90 \quad R_3 = .95 \quad R_4 = .95$$

What is the system reliability?

Solution

System reliability is often drawn schematically, as shown in Figure 11.6a. The system reliability is the reliability of one component that would be equivalent to the individual components. It could be shown as the block diagram in Figure 11.6b.

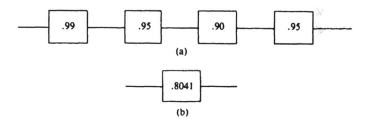

The system reliability is calculated as follows:

$$R_s = (.99)(.95)(.90)(.95) = .8041$$

Figure 11.6   Reliability Block Diagram

Parallel Reliability

Many products are designed with backup systems or backup components. The use of backup components usually makes the product more expensive, but it does improve the reliability. When a backup is present, the product will continue to function as intended as long as either the original or the backup components function as intended. *A parallel system is one that maintains reliability as long as all or any of the parallel components continue to operate.* Because of the backup nature of the system design, the overall reliability of a parallel configuration is greater than the reliability of any individual component.

To calculate the reliability of a parallel system, it is necessary to return again to the laws of probability. If two components, 1 and 2, are arranged in parallel, the system will operate:

- If component 1 operates as intended, or
- If component 2 operates as intended, or
- If both components 1 and 2 operate as intended

The addition rule for non-mutually exclusive events states that

$$P(1 \text{ or } 2) = P(1) + P(2) - P(1)P(2)$$

The reliability of two components in parallel follows the same pattern:

$$R_s = R_1 + R_2 - R_1 R_2$$

Example 11.9

Components 1 and 2 are arranged in parallel, with 1 serving as a backup system for 2. Component 1 has a reliability of .95 and component 2 has a reliability of .90. What is the system reliability?

Solution

The block diagram in Figure 11.7a illustrates the system. The diagram in Figure 11.7b shows the equivalent system reliability. The calculation of system reliability is as follows:

$$R_s = .95 + .90 - (.95)(.90) = .995$$

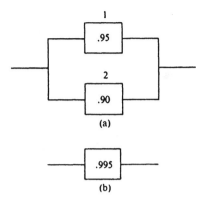

Figure 11.7   Reliability Block Diagram

The same reasoning can be expanded to three components in parallel. The system will work if

■ Component 1 works or
■ Component 2 works or
■ Component 3 works or
■ Components 1 and 2 work or
■ Components 1 and 3 work or
■ Components 2 and 3 work or
■ Components 1, 2, and 3 work

The equation for three components in a parallel system is as follows:

$$R_s = R_1 + R_2 + R_3 - R_1R_2 - R_1R_3 - R_2R_3 + R_1R_2R_3$$

Example 11.10

The Aft-Tech air taxi service owns a three engine airplane. It will function if any of the three engines functions. Motor 1 has a reliability of .99, motor has a reliability of .95, and motor 3 has a reliability of .90. What is the reliability of the airplane?

Solution

$$R_s = .99 + .95 + .90 - (.99)(.95) - (.99)(.90) - (.95)(.90)$$
$$+ (.99)(.95)(.90) = .99995$$

When there are more than three components in parallel, they may be handled as parallel systems of parallel systems. When systems get larger, a block diagram is used to show the individual components as part of the overall system.

Example 11.11

Find the reliability of the system shown in Figure 11.8a.

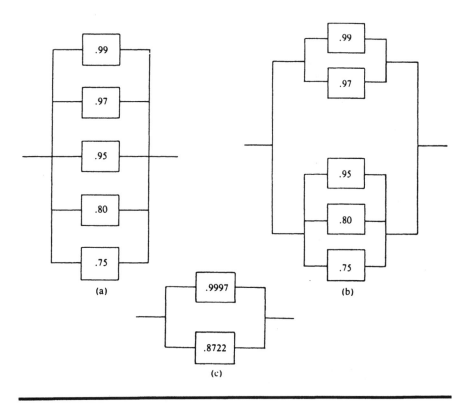

Figure 11.8   Block Diagram for Example 11.11

Solution

This system can be viewed as two small parallel systems, one of two components and one of three. This is shown in Figure 11.8b.

Each subsystem has its own reliability. The first will be designated "A" and the second "B."

$$R_A = .99 + .97 - (.99)(.97) = .9997$$

$$R_B = .95 + .80 + .75 - (.95)(.80) - (.95)(.75) - (.80)(.75) + (.95)(.80)(.75) = .8722$$

The original system of five components can now be viewed as a system of two components, A and B, in parallel. This is shown in Figure 11.8c. The reliability of this system is calculated as follows:

$$R_s = .9997 + .8722 - (.9997)(.8722) = .99996$$

If all the individual components in a parallel system or subsystem have the same reliability, then the reliability for that system can be calculated from the following special equation:

$$R_s = 1 - (1 - R_i)^n$$

where R_i is the reliability of each component and n is the number of such components in parallel.

Example 11.12

Calculate the reliability of a system of six identical components in parallel, where each component has a reliability of .8.

Solution

$$R_s = 1 - (1 - .8)^6$$
$$= 1 - (.2)^6$$
$$= 1 - .000064 = .999936$$

Some Special Cases

Most reliability systems are much more complicated than those discussed so far. Not all reliability functions follow the exponential distribution; not all systems have simple series or parallel relationships. Often designers want to know what is required in order to meet some predetermined level of reliability. The remainder of the section will illustrate, via example, some variations on what has already been developed.

Example 11.13

Determine the reliability of the system shown in Figure 11.9a.

This complex system can be subdivided into several relatively simple systems whose reliabilities can easily be calculated. These are identified as subsystems 1, 2, 3, 4, 5, and 6 in Figure 11.9b.

Solution

The reliability of each subsystem is calculated as follows:

$$R_1 = .99$$
$$R_2 = .95 + .90 - (.95)(.90) = .995$$
$$R_3 = 1 - (1 - .7)^4 = .9919$$
$$R_4 = (.9)(.9) = .81$$
$$R_5 = .81 + .8 - (.81)(.8) = .962$$
$$R_6 = .99$$

Since R_5 uses the value of R_4, the net effect of these calculations is a system as shown in Figure 11.9c. The reliability of this system is the product of the reliabilities of the individual components.

$$R_s = (.99)(.995)(.9919)(.962)(.99) = .9305$$

Example 11.14

A designer wants to ensure that the product being designed will have a system reliability of .99999. If each component has a reliability of .85, how many parallel components are required?

Solution

To determine the number of parallel components, the special redundant component relationship is solved for n rather than for R_s.

$$.99999 = 1 - (1 - .85)n$$
$$(.15)n = .00001$$
$$(n) \ln(.15) = \ln(.00001)$$
$$(n)(-1.8971) = -11.5129$$
$$n = (-11.5129)/(-1.8971) = 6.0687$$

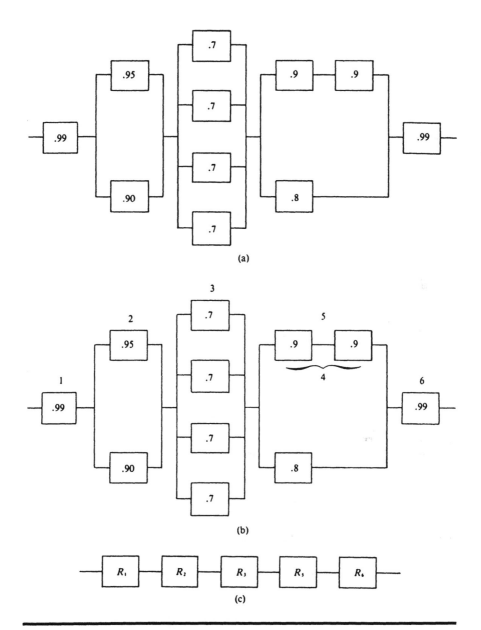

Figure 11.9   Block Diagram for Example 11.13

Thus a minimum of seven components will be required to meet the desired reliability — one original and six backup components.

Example 11.15

The mean time to failure for three components is shown in Table 11.5. The components will be arranged in a parallel configuration. What will the system reliability be for 10 hours of operation?

Table 11.5

Component	Mean
1	100
2	500
3	75

Solution

The reliability for t = 10 must be calculated for each individual component. The basic exponential reliability function is used:

$$R_1 = e^{-10/100} = .9048$$
$$R_2 = e^{-10/500} = .9802$$
$$R_3 = e^{-10/75} = .8752$$

The parallel relationship gives the following system reliability:

$$R_s = .9048 + .9802 + .8752 - (.9048)(.9802) - (.9048)(.8752)$$
$$- (.9802)(.87852) + (.9048)(.9802)(.8752) = .99976$$

Summary

Reliability of certain components and products is important. Predicting the reliability of specific components can involve complicated mathematics, but use of predicted reliabilities can greatly enhance the quality of the product by helping to ensure that the product performs consistently in the manner expected.

Practice Problems

1a. A product has a mean time between failures of 800 hours. What is its reliability for the following operation times?
   (1) 200 hours
   (2) 400 hours
   (3) 1000 hours
   (4) 800 hours

1b. A component has a failure rate of .0001 hours. What is its reliability for the following operation times?
   (1) 100,000 hours
   (2) 20,000 hours
   (3) 5,000 hours

2. A component must have a reliability of .95 for 5,000 hours of service.
   (a) What is the failure rate?
   (b) What is the MTBF?

3. A component must have a reliability of .9 for 20,000 minutes of service.
   (a) What is the failure rate?
   (b) What is the MTBF?

4. A component must have a reliability of .99 for 100 hours of service.
   (a) What is the failure rate?
   (b) What is the MTTF?

5. A testing lab tested ten products to determine the maximum service life. These products lasted 100, 110, 120, 140, 100, 90, 110, 120, and 120 hours, respectively. What is the MTTF?

6. The table below lists times between failures for some components that were life tested by a testing lab. The product is expected to be used for 30,000 miles. What is the reliability?

Frequency	Life (Miles)
5	40,000
15	45,000
25	50,000
30	55,000
50	60,000
25	65,000
150	

7. Rocket system backup units for moon blastoff each have a component reliability of .9. It is desired to give the system of backup units a reliability of .99999 when each backup unit functions independently. How many backup systems should be used?

8. Shown below are four systems. Determine the reliability of each.

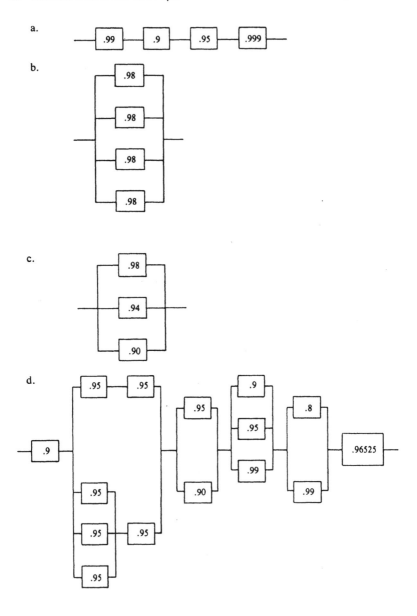

Appendix

Answers to Selected Problems

Chapter 3

Number	Answer
2	0.625
4	0.25
6	0.000455
8	0.000003283
10	0.6
12	0.6
14	0.16
16	0.1444
18	0.3

Chapter 4

Number	Answer
2	0.448
4	0.4221
6	0.0106
8	0.0017
10	0.0122
12	0.0731
14	0.1519

16	0.715
18	0.1606
20	0.003
22	0.5419
24	0.2033
26	0.7967
28	0.2033
30	0.069
32	0.0656
34	0.0475
36	0.0558
38	0.0011
40	0.2953
42	0.9835
44	0.0178
46	299 and 201
50	167
53a	0.1894
53b	0.0102
53c	0.6808
53d	0.9984
53e	0.4236
58a	0.4074
58b	0.6596

Chapter 5

Number	Answer	
1	14.13	11.27
3	414.35	409.65
7	0.252	0
10	0.516	0.324
13	2.106	1.316
17	0.476	0.384
21	0.147	0
22	0.2245	0.089
24	2.324	2.079

Chapter 6

Number	Test Statistic
1	z test = 1.20
3	t test = 2.19

5	t test = .624
7	t test = .8802
10	t test = 2.81
13	F test = 7.596
16	t test = 1.80

Chapter 7

Number	Answer
2	$UCL_{\bar{x}} = 131.06$
	$LCL_{\bar{x}} = 118.94$
	$UCL_R = 32.43$
	$LCL_R = 3.21$
	$UCL_x = 143$
	$LCL_x = 107$
4	$UCL_{\bar{x}} = 76.38$
	$LCL_{\bar{x}} = 67.62$
	$UCL_R = 13.68$
	$LCL_R = 0$
	$UCL_x = 80.74$
	$LCL_x = 63.26$
7	$UCL_{\bar{x}} = 80.04$
	$LCL_{\bar{x}} = 69.96$
	$UCL_R = 23.04$
	$LCL_R = .96$
	$UCL_x = 88.31$
	$LCL_x = 61.69$
10	$UCL_c = 13.35$
	$LCL_c = 0$
16	Upper = 21.03
	Lower = 4.77

Chapter 8

Number	Answer
2	UNTL = 56.21
	LNTL = 43.79
4	10.96% or 66
6	10.24%
9	$C_{pk} = .54$
13	$C_p = 1.25$ $C_{pk} = 1.25$
15	579.3/1000
	579,300/1,000,000

Chapter 10

Number	Answer	
1	0.17	
3	0.265	
5	n = 216	c = 5
8	n = 200	c = 4
11	0.3684	
13a	0.0211	
13b	0.0089	
16a	77.15	
16b	66.95	
16c	154.34	
17a	645.96	
18a	763.13	

Chapter 11

Number	Answer
1a(1)	0.7788
1b(1)	0.00005
2a	0.00001026
3b	189824.4
4a	0.000101
5	112.22
7	5
8a	0.8456
8c	0.99988

Table A Binomial Distribution

		p								
n	*x*	*0.1*	*0.2*	*0.3*	*0.4*	*0.5*	*0.6*	*0.7*	*0.8*	*0.9*
1	0	0.9000	0.8000	0.7000	0.6000	0.5000	0.4000	0.3000	0.2000	0.1000
1	1	0.1000	0.2000	0.3000	0.4000	0.5000	0.6000	0.7000	0.8000	0.9000
2	0	0.8100	0.6400	0.4900	0.3600	0.2500	0.1600	0.0900	0.0400	0.0100
2	1	0.1800	0.3200	0.4200	0.4800	0.5000	0.4800	0.4200	0.3200	0.1800
2	2	0.0100	0.0400	0.0900	0.1600	0.2500	0.3600	0.4900	0.6400	0.8100
3	0	0.7290	0.5120	0.3430	0.2160	0.1250	0.0640	0.0270	0.0080	0.0010
3	1	0.2430	0.3840	0.4410	0.4320	0.3750	0.2880	0.1890	0.0960	0.0270
3	2	0.0270	0.0960	0.1890	0.2880	0.3750	0.4320	0.4410	0.3840	0.2430
3	3	0.0010	0.0080	0.0270	0.0640	0.1250	0.2160	0.3430	0.5120	0.7290
4	0	0.6561	0.4096	0.2401	0.1296	0.0625	0.0256	0.0081	0.0016	0.0001
4	1	0.2916	0.4096	0.4116	0.3456	0.2500	0.1536	0.0756	0.0256	0.0036
4	2	0.0486	0.1536	0.2646	0.3456	0.3750	0.3456	0.2646	0.1536	0.0486
4	3	0.0036	0.0256	0.0756	0.1536	0.2500	0.3456	0.4116	0.4096	0.2916
4	4	0.0001	0.0016	0.0081	0.0256	0.0625	0.1296	0.2401	0.4096	0.6561
5	0	0.5905	0.3277	0.1681	0.0778	0.0313	0.0102	0.0024	0.0003	0.0000
5	1	0.3281	0.4096	0.3602	0.2592	0.1563	0.0768	0.0284	0.0064	0.0004
5	2	0.0729	0.2048	0.3087	0.3456	0.3125	0.2304	0.1323	0.0512	0.0081
5	3	0.0081	0.0512	0.1323	0.2304	0.3125	0.3456	0.3087	0.2048	0.0729
5	4	0.0005	0.0064	0.0284	0.0768	0.1563	0.2592	0.3602	0.4096	0.3281
5	5	0.0000	0.0003	0.0024	0.0102	0.0313	0.0778	0.1681	0.3277	0.5905
6	0	0.5314	0.2621	0.1176	0.0467	0.0156	0.0041	0.0007	0.0001	0.0000
6	1	0.3543	0.3932	0.3025	0.1866	0.0938	0.0369	0.0102	0.0015	0.0001
6	2	0.0984	0.2458	0.3241	0.3110	0.2344	0.1382	0.0595	0.0154	0.0012
6	3	0.0146	0.0819	0.1852	0.2765	0.3125	0.2765	0.1852	0.0819	0.0146
6	4	0.0012	0.0154	0.0595	0.1382	0.2344	0.3110	0.3241	0.2458	0.0984
6	5	0.0001	0.0015	0.0102	0.0369	0.0938	0.1866	0.3025	0.3932	0.3543
6	6	0.0000	0.0001	0.0007	0.0041	0.0156	0.0467	0.1176	0.2621	0.5314
7	0	0.4783	0.2097	0.0824	0.0280	0.0078	0.0016	0.0002	0.0000	0.0000
7	1	0.3720	0.3670	0.2471	0.1306	0.0547	0.0172	0.0036	0.0004	0.0000
7	2	0.1240	0.2753	0.3177	0.2613	0.1641	0.0774	0250	0.0043	0.0002
7	3	0.0230	0.1147	0.2269	0.2903	0.2734	0.1935	0.0972	0.0287	0.0026
7	4	0.0026	0.0287	0.0972	0.1935	0.2734	0.2903	0.2269	0.1147	0.0230
7	5	0.0002	0.0043	0.0250	0.0774	0.1641	0.2613	0.3177	0.2753	0.1240
7	6	0.0000	0.0004	0.0036	0.0172	0.0547	0.1306	0.2471	0.3670	0.3720
7	7	0.0000	0.0000	0.0002	0.0016	0.0078	0.0280	0.0824	0.2097	0.4783

Table A　Binomial Distribution (continued)

n	x	\multicolumn{9}{c}{p}
		0.1	0.2	0.3	0.4	0.5	0.6	0.7	0.8	0.9
8	0	0.4305	0.1678	0.0576	0.0168	0.0039	0.0007	0.0001	0.0000	0.0000
8	1	0.3826	0.3355	0.1977	0.0896	0.0313	0.0079	0.0012	0.0001	0.0000
8	2	0.1488	0.2936	0.2965	0.2090	0.1094	0.0413	0.0100	0.0011	0.0000
8	3	0.0331	0.1468	0.2541	0.2787	0.2188	0.1239	0.0467	0.0092	0.0004
8	4	0.0046	0.0459	0.1361	0.2322	0.2734	0.2322	0.1361	0.0459	0.0046
8	5	0.0004	0.0092	0.0467	0.1239	0.2188	0.2787	0.2541	0.1468	0.0331
8	6	0.0000	0.0011	0.0100	0.0413	0.1094	0.2090	0.2965	0.2936	0.1488
8	7	0.0000	0.0001	0.0012	0.0079	0.0313	0.0896	0.1977	0.3355	0.3826
8	8	0.0000	0.0000	0.0001	0.0007	0.0039	0.0168	0.0576	0.1678	0.4305
9	0	0.3874	0.1342	0.0404	0.0101	0.0020	0.0003	0.0000	0.0000	0.0000
9	1	0.3874	0.3020	0.1556	0.0605	0.0176	0.0035	0.0004	0.0000	0.0000
9	2	0.1722	0.3020	0.2668	0.1612	0.0703	0.0212	0.0039	0.0003	0.0000
9	3	0.0446	0.1762	0.2668	0.2508	0.1641	0.0743	0.0210	0.0028	0.0001
9	4	0.0074	0.0661	0.1715	0.2508	0.2461	0.1672	0.0735	0.0165	0.0008
9	5	0.0008	0.0165	0.0735	0.1672	0.2461	0.2508	0.1715	0.0661	0.0074
9	6	0.0001	0.0028	0.0210	0.0743	0.1641	0.2508	0.2668	0.1762	0.0446
9	7	0.0000	0.0003	0.0039	0.0212	0.0703	0.1612	0.2668	0.3020	0.1722
9	8	0.0000	0.0000	0.0004	0.0035	0.0176	0.0605	0.1556	0.3020	0.3874
9	9	0.0000	0.0000	0.0000	0.0003	0.0020	0.0101	0.0404	0.1342	0.3874
10	0	0.3487	0.1074	0.0282	0.0060	0.0010	0.0001	0.0000	0.0000	0.0000
10	1	0.3874	0.2684	0.1211	0.0403	0.0098	0.0016	0.0001	0.0000	0.0000
10	2	0.1937	0.3020	0.2335	0.1209	0.0439	0.0106	0.0014	0.0001	0.0000
10	3	0.0574	0.2013	0.2668	0.2150	0.1172	0.0425	0.0090	0.0008	0.0000
10	4	0.0112	0.0881	0.2001	0.2508	0.2051	0.1115	0.0368	0.0055	0.0001
10	5	0.0015	0.0264	0.1029	0.2007	0.2461	0.2007	0.1029	0.0264	0.0015
10	6	0.0001	0.0055	0.0368	0.1115	0.2051	0.2508	0.2001	0.0881	0.0112
10	7	0.0000	0.0008	0.0090	0.0425	0.1172	0.2150	0.2668	0.2013	0.0574
10	8	0.0000	0.0001	0.0014	0.0106	0.0439	0.1209	0.2335	0.3020	0.1937
10	9	0.0000	0.0000	0.0001	0.0016	0.0098	0.0403	0.1211	0.2684	0.3874
10	10	0.0000	0.0000	0.0000	0.0001	0.0010	0.0060	0.0282	0.1074	0.3487
11	0	0.3138	0.0859	0.0198	0.0036	0.0005	0.0000	0.0000	0.0000	0.0000
11	1	0.3835	0.2362	0.0932	0.0266	0.0054	0.0007	0.0000	0.0000	0.0000
11	2	0.2131	0.2953	0.1998	0.0887	0.0269	0.0052	0.0005	0.0000	0.0000
11	3	0.0710	0.2215	0.2568	0.1774	0.0806	0.0234	0.0037	0.0002	0.0000
11	4	0.0158	0.1107	0.2201	0.2365	0.1611	0.0701	0.0173	0.0017	0.0000
11	5	0.0025	0.0388	0.1312	0.2207	0.2256	0.1471	0.0566	0.0097	0.0003
11	6	0.0003	0.0097	0.0566	0.1471	0.2256	0.2207	0.1321	0.0388	0.0025
11	7	0.0000	0.0017	0.0173	0.0701	0.1611	0.2365	0.2201	0.1107	0.0158

Table A   Binomial Distribution (continued)

n	x	0.1	0.2	0.3	0.4	0.5	0.6	0.7	0.8	0.9
11	8	0.0000	0.0002	0.0037	0.0234	0.0806	0.1774	0.2568	0.2215	0.0710
11	9	0.0000	0.0000	0.0005	0.0052	0.0269	0.0887	0.1998	0.2953	0.2131
11	10	0.0000	0.0000	0.0000	0.0007	0.0054	0.0266	0.0932	0.2362	0.3835
11	11	0.0000	0.0000	0.0000	0.0000	0.0005	0.0036	0.0198	0.0859	0.3138
12	0	0.2824	0.0687	0.0138	0.0022	0.0002	0.0000	0.0000	0.0000	0.0000
12	1	0.3766	0.2062	0.0712	0.0174	0.0029	0.0003	0.0000	0.0000	0.0000
12	2	0.2301	0.2835	0.1678	0.0639	0.0161	0.0025	0.0002	0.0000	0.0000
12	3	0.0852	0.2362	0.2397	0.1419	0.0537	0.0125	0.0015	0.0001	0.0000
12	4	0.0213	0.1329	0.2311	0.2128	0.1208	0.0420	0.0078	0.0005	0.0000
12	5	0.0038	0.0532	0.1585	0.2270	0.1934	0.1009	0.0291	0.0033	0.0000
12	6	0.0005	0.0155	0.0792	0.1766	0.2256	0.1766	0.0792	0.0155	0.0005
12	7	0.0000	0.0033	0.0291	0.1009	0.1934	0.2270	0.1585	0.0532	0.0038
12	8	0.0000	0.0005	0.0078	0.0420	0.1208	0.2128	0.2311	0.1329	0.0213
12	9	0.0000	0.0001	0.0015	0.0125	0.0537	0.1419	0.2397	0.2362	0.0852
12	10	0.0000	0.0000	0.0002	0.0025	0.0161	0.0639	0.1678	0.2835	0.2301
12	11	0.0000	0.0000	0.0000	0.0003	0.0029	0.0174	0.0712	0.2062	0.3766
12	12	0.0000	0.0000	0.0000	0.0000	0.0002	0.0022	0.0138	0.0687	0.2824
13	0	0.2542	0.0550	0.0097	0.001	0.0001	0.0000	0.0000	0.0000	0.0000
13	1	0.3672	0.1787	0.0540	0.0113	0.0016	0.0001	0.0000	0.0000	0.0000
13	2	0.2448	0.2680	0.1388	0.0453	0.0095	0.0012	0.0001	0.0000	0.0000
13	3	0.0997	0.2457	0.2181	0.1107	0.0349	0.0065	0.0006	0.0000	0.0000
13	4	0.0277	0.1535	0.2337	0.1845	0.0873	0.0243	0.0034	0.0001	0.0000
13	5	0.0055	0.0691	0.1803	0.2214	0.1571	0.0656	0.0142	0.0011	0.0000
13	6	0.0008	0.0230	0.1030	0.1968	0.2095	0.1312	0.0442	0.0058	0.0001
13	7	0.0001	0.0058	0.0442	0.1312	0.2095	0.1968	0.1030	0.0230	0.0008
13	8	0.0000	0.0011	0.0142	0.0656	0.1571	0.2214	0.1803	0.0691	0.0055
13	9	0.0000	0.0001	0.0034	0.0243	0.0873	0.1845	0.2337	0.1535	0.0277
13	10	0.0000	0.0000	0.0006	0.0065	0.0349	0.1107	0.2181	0.2457	0.0997
13	11	0.0000	0.0000	0.0001	0.0012	0.0095	0.0453	0.1388	0.2680	0.2448
13	12	0.0000	0.0000	0.0000	0.0001	0.0016	0.0113	0.0540	0.1787	0.3672
13	13	0.0000	0.0000	0.0000	0.0000	0.0001	0.0013	0.0097	0.0550	0.2542
14	0	0.2288	0.0440	0.0068	0.0008	0.0001	0.0000	0.0000	0.0000	0.0000
14	1	0.3559	0.1539	0.0407	0.0073	0.0009	0.0001	0.0000	0.0000	0.0000
14	2	0.2570	0.2501	0.1134	0.0317	0.0056	0.0005	0.0000	0.0000	0.0000
14	3	0.1142	0.2501	0.1943	0.0845	0.0222	0.0033	0.0002	0.0000	0.0000
14	4	0.0349	0.1720	0.2290	0.1549	0.0611	0.0136	0.0014	0.0000	0.0000
14	5	0.0078	0.0860	0.1963	0.2066	0.1222	0.0408	0.0066	0.0003	0.0000
14	6	0.0013	0.0322	0.1262	0.2066	0.1833	0.0918	0.0232	0.0020	0.0000

Table A   Binomial Distribution (continued)

n	x	0.1	0.2	0.3	0.4	0.5	0.6	0.7	0.8	0.9
14	7	0.0002	0.0092	0.0618	0.1574	0.2095	0.1574	0.0618	0.0092	0.0002
14	8	0.0000	0.0020	0.0232	0.0918	0.1833	0.2066	0.1262	0.0322	0.0013
14	9	0.0000	0.0003	0.0066	0.0408	0.1222	0.2066	0.1963	0.0860	0.0078
14	10	0.0000	0.0000	0.0014	0.0136	0.0611	0.1549	0.2290	0.1720	0.0349
14	11	0.0000	0.0000	0.0002	0.0033	0.0222	0.0845	0.1943	0.2501	0.1142
14	12	0.0000	0.0000	0.0000	0.0005	0.0056	0.0317	0.1134	0.2501	0.2570
14	13	0.0000	0.0000	0.0000	0.0001	0.0009	0.0073	0.0407	0.1539	0.3559
14	14	0.0000	0.0000	0.0000	0.0000	0.0001	0.0008	0.0068	0.0440	0.2288
15	0	0.2059	0.0352	0.0047	0.0005	0.0000	0.0000	0.0000	0.0000	0.0000
15	1	0.3432	0.1319	0.0305	0.0047	0.0005	0.0000	0.0000	0.0000	0.0000
15	2	0.2669	0.2309	0.0916	0.0219	0.0032	0.0003	0.0000	0.0000	0.0000
15	3	0.1285	0.2501	0.1700	0.0634	0.0139	0.0016	0.0001	0.0000	0.0000
15	4	0.0428	0.1876	0.2186	0.1268	0.0417	0.0074	0.0006	0.0000	0.0000
15	5	0.0105	0.1032	0.2061	0.1859	0.0916	0.0245	0.0030	0.0001	0.0000
15	6	0.0019	0.0430	0.1472	0.2066	0.1527	0.0612	0.0116	0.0007	0.0000
15	7	0.0003	0.0138	0.0811	0.1771	0.1964	0.1181	0.0348	0.0035	0.0000
15	8	0.0000	0.0035	0.0348	0.1181	0.1964	0.1771	0.0811	0.0018	0.0003
15	9	0.0000	0.0007	0.0116	0.0612	0.1527	0.2066	0.1472	0.0430	0.0019
15	10	0.0000	0.0001	0.0030	0.0245	0.0916	0.1859	0.2061	0.1032	0.0105
15	11	0.0000	0.0000	0.0006	0.0074	0.0417	0.1268	0.2186	0.1876	0.0428
15	12	0.0000	0.0000	0.0001	0.0016	0.0139	0.0634	0.1700	0.2501	0.1285
15	13	0.0000	0.0000	0.0000	0.0003	0.0032	0.0219	0.0916	0.2309	0.2669
15	14	0.0000	0.0000	0.0000	0.0000	0.0005	0.0047	0.0305	0.1319	0.0432
15	15	0.0000	0.0000	0.0000	0.0000	0.0000	0.0004	0.0047	0.0352	0.2059
20	0	0.1216	0.0115	0.0008	0.0000	0.0000	0.0000	0.0000	0.0000	0.0000
20	1	0.2702	0.0576	0.0068	0.0005	0.0000	0.0000	0.0000	0.0000	0.0000
20	2	0.2852	0.1369	0.0278	0.0031	0.0002	0.0000	0.0000	0.0000	0.0000
20	3	0.1901	0.2054	0.0716	0.0123	0.0011	0.0000	0.0000	0.0000	0.0000
20	4	0.0898	0.2182	0.1304	0.0350	0.0046	0.0003	0.0000	0.0000	0.0000
20	5	0.0319	0.1746	0.1789	0.0746	0.0148	0.0013	0.0000	0.0000	0.0000
20	6	0.0089	0.1091	0.1916	0.1244	0.0370	0.0049	0.0002	0.0000	0.0000
20	7	0.0020	0.0545	0.1643	0.1659	0.0739	0.0146	0.0010	0.0000	0.0000
20	8	0.0004	0.0222	0.1144	0.1797	0.1201	0.0355	0.0039	0.0001	0.0000
20	9	0.0001	0.0074	0.0654	0.1597	0.1602	0.0710	0.0120	0.0005	0.0000
20	10	0.0000	0.0020	0.0308	0.1171	0.1762	0.1171	0.0308	0.0020	0.0000
20	11	0.0000	0.0005	0.0120	0.0710	0.1602	0.1597	0.0654	0.0074	0.0001
20	12	0.0000	0.0001	0.0039	0.0355	0.1210	0.1797	0.1144	0.0222	0.0004
20	13	0.0000	0.0000	0.0010	0.0146	0.0739	0.1659	0.1643	0.0545	0.0020
20	14	0.0000	0.0000	0.0002	0.0049	0.0370	0.1244	0.1916	0.1091	0.0089

Table A Binomial Distribution (continued)

n	x	0.1	0.2	0.3	0.4	0.5	0.6	0.7	0.8	0.9
20	15	0.0000	0.0000	0.0000	0.0013	0.0148	0.0746	0.1789	0.1746	0.0319
20	16	0.0000	0.0000	0.0000	0.0003	0.0046	0.0350	0.1304	0.2182	0.0898
20	17	0.0000	0.0000	0.0000	0.0000	0.0011	0.0123	0.0716	0.2054	0.1901
20	18	0.0000	0.0000	0.0000	0.0000	0.0002	0.0031	0.0278	0.1369	0.2852
20	19	0.0000	0.0000	0.0000	0.0000	0.0000	0.0005	0.0068	0.0576	0.2702
20	20	0.0000	0.0000	0.0000	0.0000	0.0000	0.0000	0.0008	0.0115	0.1216
25	0	0.0718	0.0038	0.0001	0.0000	0.0000	0.0000	0.0000	0.0000	0.0000
25	1	0.1994	0.0236	0.0014	0.0000	0.0000	0.0000	0.0000	0.0000	0.0000
25	2	0.2659	0.0708	0.0074	0.0004	0.0000	0.0000	0.0000	0.0000	0.0000
25	3	0.2265	0.1358	0.0243	0.0019	0.0001	0.0000	0.0000	0.0000	0.0000
25	4	0.1384	0.1867	0.0572	0.0071	0.0004	0.0000	0.0000	0.0000	0.0000
25	5	0.0646	0.1960	0.1030	0.0199	0.0016	0.0000	0.0000	0.0000	0.0000
25	6	0.0239	0.1633	0.1472	0.0442	0.0053	0.0002	0.0000	0.0000	0.0000
25	7	0.0072	0.1108	0.1712	0.0800	0.0143	0.0009	0.0000	0.0000	0.0000
25	8	0.0018	0.0623	0.1651	0.1200	0.0322	0.0031	0.0001	0.0000	0.0000
25	9	0.0004	0.0294	0.1336	0.1511	0.0609	0.0088	0.0004	0.0000	0.0000
25	10	0.0001	0.0118	0.0916	0.1612	0.0974	0.0212	0.0013	0.0000	0.0000
25	11	0.0000	0.0040	0.0536	0.1465	0.1328	0.0434	0.0042	0.0001	0.0000
25	12	0.0000	0.0012	0.0268	0.1140	0.1550	0.0760	0.0115	0.0003	0.0000
25	13	0.0000	0.0003	0.0115	0.0760	0.1550	0.1140	0.0268	0.0012	0.0000
25	14	0.0000	0.0001	0.0042	0.0434	0.1328	0.1465	0.0536	0.0040	0.0000
25	15	0.0000	0.0000	0.0013	0.0212	0.0974	0.1612	0.0916	0.0118	0.0001
25	16	0.0000	0.0000	0.0004	0.0088	0.0609	0.1511	0.1336	0.0294	0.0004
25	17	0.0000	0.0000	0.0001	0.0031	0.0322	0.1200	0.1651	0.0623	0.0018
25	18	0.0000	0.0000	0.0000	0.0009	0.0143	0.0800	0.1712	0.1108	0.0072
25	19	0.0000	0.0000	0.0000	0.0002	0.0053	0.0442	0.1472	0.1633	0.0239
25	20	0.0000	0.0000	0.0000	0.0000	0.0016	0.0199	0.1030	0.1960	0.0646
25	21	0.0000	0.0000	0.0000	0.0000	0.0004	0.0071	0.0572	0.1867	0.1384
25	22	0.0000	0.0000	0.0000	0.0000	0.0001	0.0019	0.0243	0.1358	0.2265
25	23	0.0000	0.0000	0.0000	0.0000	0.0000	0.0004	0.0074	0.0708	0.2659
25	24	0.0000	0.0000	0.0000	0.0000	0.0000	0.0000	0.0014	0.0326	0.1994
25	25	0.0000	0.0000	0.0000	0.0000	0.0000	0.0000	0.0001	0.0038	0.0708

Table B   Exact Poisson Probabilities

np	0	1	2	3	4	5	6	7	8	9	10	11	12	13
0.02	0.9802	0.0196	0.0002	0.0000	0.0000	0.0000	0.0000	0.0000	0.0000	0.0000	0.0000	0.0000	0.0000	0.0000
0.04	0.9608	0.0384	0.0008	0.0000	0.0000	0.0000	0.0000	0.0000	0.0000	0.0000	0.0000	0.0000	0.0000	0.0000
0.06	0.9418	0.565	0.0017	0.0000	0.0000	0.0000	0.0000	0.0000	0.0000	0.0000	0.0000	0.0000	0.0000	0.0000
0.08	0.9231	0.0738	0.0030	0.0001	0.0000	0.0000	0.0000	0.0000	0.0000	0.0000	0.0000	0.0000	0.0000	0.0000
0.10	0.9048	0.0905	0.0045	0.0002	0.0000	0.0000	0.0000	0.0000	0.0000	0.0000	0.0000	0.0000	0.0000	0.0000
0.12	0.8869	0.1064	0.0064	0.0003	0.0000	0.0000	0.0000	0.0000	0.0000	0.0000	0.0000	0.0000	0.0000	0.0000
0.14	0.8694	0.1217	0.0085	0.0004	0.0000	0.0000	0.0000	0.0000	0.0000	0.0000	0.0000	0.0000	0.0000	0.0000
0.16	0.8521	0.1363	0.0109	0.0006	0.0000	0.0000	0.0000	0.0000	0.0000	0.0000	0.0000	0.0000	0.0000	0.0000
0.18	0.8353	0.1503	0.0135	0.0008	0.0000	0.0000	0.0000	0.0000	0.0000	0.0000	0.0000	0.0000	0.0000	0.0000
0.20	0.8187	0.1637	0.0164	0.0011	0.0001	0.0000	0.0000	0.0000	0.0000	0.0000	0.0000	0.0000	0.0000	0.0000
0.22	0.8025	0.1766	0.0194	0.0014	0.0001	0.0000	0.0000	0.0000	0.0000	0.0000	0.0000	0.0000	0.0000	0.0000
0.24	0.7866	0.1888	0.0227	0.0018	0.0001	0.0000	0.0000	0.0000	0.0000	0.0000	0.0000	0.0000	0.0000	0.0000
0.26	0.7711	0.2005	0.0261	0.0023	0.0001	0.0000	0.0000	0.0000	0.0000	0.0000	0.0000	0.0000	0.0000	0.0000
0.28	0.7558	0.2116	0.0296	0.0028	0.0002	0.0000	0.0000	0.0000	0.0000	0.0000	0.0000	0.0000	0.0000	0.0000
0.30	0.7408	0.2222	0.0333	0.0033	0.0003	0.0000	0.0000	0.0000	0.0000	0.0000	0.0000	0.0000	0.0000	0.0000
0.32	0.7261	0.2324	0.0372	0.0040	0.0003	0.0000	0.0000	0.0000	0.0000	0.0000	0.0000	0.0000	0.0000	0.0000
0.34	0.7118	0.2420	0.0411	0.0047	0.0004	0.0000	0.0000	0.0000	0.0000	0.0000	0.0000	0.0000	0.0000	0.0000
0.36	0.6977	0.2512	0.0452	0.0054	0.0005	0.0000	0.0000	0.0000	0.0000	0.0000	0.0000	0.0000	0.0000	0.0000
0.38	0.6839	0.2599	0.0494	0.0063	0.0006	0.0000	0.0000	0.0000	0.0000	0.0000	0.0000	0.0000	0.0000	0.0000
0.40	0.6703	0.2681	0.0536	0.0072	0.0007	0.0001	0.0000	0.0000	0.0000	0.0000	0.0000	0.0000	0.0000	0.0000
0.42	0.6570	0.2760	0.0580	0.0081	0.0009	0.0001	0.0000	0.0000	0.0000	0.0000	0.0000	0.0000	0.0000	0.0000
0.44	0.6440	0.2834	0.0623	0.0091	0.0010	0.0001	0.0000	0.0000	0.0000	0.0000	0.0000	0.0000	0.0000	0.0000
0.46	0.6313	0.2904	0.0668	0.0102	0.0012	0.0001	0.0000	0.0000	0.0000	0.0000	0.0000	0.0000	0.0000	0.0000
0.48	0.6188	0.2970	0.0713	0.0114	0.0014	0.0001	0.0000	0.0000	0.0000	0.0000	0.0000	0.0000	0.0000	0.0000
0.50	0.6065	0.3033	0.0758	0.0126	0.0016	0.0002	0.0000	0.0000	0.0000	0.0000	0.0000	0.0000	0.0000	0.0000
0.55	0.5769	0.3173	0.0873	0.0160	0.0022	0.0002	0.0000	0.0000	0.0000	0.0000	0.0000	0.0000	0.0000	0.0000
0.60	0.5488	0.3293	0.0988	0.0198	0.0030	0.0004	0.0000	0.0000	0.0000	0.0000	0.0000	0.0000	0.0000	0.0000

0.65	0.5220	0.3393	0.1103	0.0239	0.0039	0.0005	0.0001	0.0000	0.0000	0.0000	0.0000	0.0000	0.0000	0.0000
0.70	0.4966	0.3476	0.1217	0.0284	0.0050	0.0007	0.0001	0.0000	0.0000	0.0000	0.0000	0.0000	0.0000	0.0000
0.75	0.4724	0.3543	0.1329	0.0332	0.0062	0.0009	0.0001	0.0000	0.0000	0.0000	0.0000	0.0000	0.0000	0.0000
0.80	0.4493	0.3595	0.1438	0.0383	0.0077	0.0012	0.0002	0.0000	0.0000	0.0000	0.0000	0.0000	0.0000	0.0000
0.85	0.4274	0.3633	0.1544	0.0437	0.0093	0.0016	0.0002	0.0000	0.0000	0.0000	0.0000	0.0000	0.0000	0.0000
0.90	0.4066	0.3659	0.1647	0.0494	0.0111	0.0020	0.0003	0.0000	0.0000	0.0000	0.0000	0.0000	0.0000	0.0000
0.95	0.3867	0.3674	0.1745	0.0553	0.0131	0.0025	0.0004	0.0001	0.0000	0.0000	0.0000	0.0000	0.0000	0.0000
1.00	0.3679	0.3679	0.1839	0.0613	0.0153	0.0031	0.0005	0.0001	0.0000	0.0000	0.0000	0.0000	0.0000	0.0000
1.05	0.3499	0.3674	0.1929	0.0675	0.0177	0.0037	0.0007	0.0001	0.0000	0.0000	0.0000	0.0000	0.0000	0.0000
1.10	0.3329	0.3662	0.2014	0.0738	0.0203	0.0045	0.0008	0.0001	0.0000	0.0000	0.0000	0.0000	0.0000	0.0000
1.15	0.3166	0.3641	0.2094	0.0803	0.0231	0.0053	0.0010	0.0002	0.0000	0.0000	0.0000	0.0000	0.0000	0.0000
1.20	0.3012	0.3614	0.2169	0.0867	0.0260	0.0062	0.0012	0.0002	0.0000	0.0000	0.0000	0.0000	0.0000	0.0000
1.25	0.2865	0.3581	0.2238	0.0933	0.0291	0.0073	0.0015	0.0003	0.0000	0.0000	0.0000	0.0000	0.0000	0.0000
1.30	0.2725	0.3543	0.2303	0.0998	0.0324	0.0084	0.0018	0.0003	0.0000	0.0000	0.0000	0.0000	0.0000	0.0000
1.35	0.2592	0.3500	0.2362	0.1063	0.0359	0.0097	0.0022	0.0004	0.0001	0.0000	0.0000	0.0000	0.0000	0.0000
1.40	0.2466	0.3452	0.2417	0.1128	0.0395	0.0111	0.0026	0.0005	0.0001	0.0000	0.0000	0.0000	0.0000	0.0000
1.45	0.2346	0.3401	0.2466	0.1192	0.0432	0.0125	0.0030	0.0006	0.0001	0.0000	0.0000	0.0000	0.0000	0.0000
1.50	0.2231	0.3347	0.2510	0.1255	0.0471	0.0141	0.0035	0.0008	0.0002	0.0000	0.0000	0.0000	0.0000	0.0000
1.55	0.2122	0.3290	0.2550	0.1317	0.0510	0.0158	0.0041	0.0009	0.0002	0.0000	0.0000	0.0000	0.0000	0.0000
1.60	0.2019	0.3230	0.2584	0.1378	0.0551	0.0176	0.0047	0.0011	0.0002	0.0000	0.0000	0.0000	0.0000	0.0000
1.65	0.1920	0.3169	0.2614	0.1438	0.0593	0.0196	0.0054	0.0013	0.0003	0.0000	0.0000	0.0000	0.0000	0.0000
1.70	0.1827	0.3106	0.2640	0.1496	0.0636	0.0216	0.0061	0.0015	0.0003	0.0000	0.0000	0.0000	0.0000	0.0000
1.75	0.1738	0.3041	0.2661	0.1552	0.0679	0.0238	0.0069	0.0017	0.0004	0.0000	0.0000	0.0000	0.0000	0.0000
1.80	0.1653	0.2975	0.2678	0.1607	0.0723	0.0260	0.0078	0.0020	0.0005	0.0001	0.0000	0.0000	0.0000	0.0000
1.85	0.1572	0.2909	0.2691	0.1659	0.0767	0.0284	0.0088	0.0023	0.0005	0.0001	0.0000	0.0000	0.0000	0.0000
1.90	0.1496	0.2842	0.2700	0.1710	0.0812	0.0309	0.0098	0.0027	0.0006	0.0001	0.0000	0.0000	0.0000	0.0000
1.95	0.1423	0.2774	0.2705	0.1758	0.0857	0.0334	0.0109	0.0030	0.0007	0.0002	0.0000	0.0000	0.0000	0.0000
2.00	0.1353	0.2707	0.2707	0.1804	0.0902	0.0361	0.0120	0.0034	0.0009	0.0002	0.0000	0.0000	0.0000	0.0000
2.05	0.1287	0.2639	0.2705	0.1848	0.0947	0.0388	0.0133	0.0039	0.0010	0.0002	0.0000	0.0000	0.0000	0.0000
2.10	0.1225	0.2572	0.2700	0.1890	0.0992	0.0417	0.0146	0.0044	0.0011	0.0003	0.0001	0.0000	0.0000	0.0000
2.15	0.1165	0.2504	0.2692	0.1929	0.1037	0.0446	0.0160	0.0049	0.0013	0.0003	0.0001	0.0000	0.0000	0.0000

Table B   Exact Poisson Probabilities (continued)

np	0	1	2	3	4	5	6	7	8	9	10	11	12	13
2.20	0.1108	0.2438	0.2681	0.1966	0.1082	0.0476	0.0174	0.0055	0.0015	0.0004	0.0001	0.0000	0.0000	0.0000
2.25	0.1054	0.2371	0.2668	0.2001	0.1126	0.0506	0.0190	0.0061	0.0017	0.0004	0.0001	0.0000	0.0000	0.0000
2.30	0.1003	0.2306	0.2652	0.2033	0.1169	0.0538	0.0206	0.0068	0.0019	0.0005	0.0001	0.0000	0.0000	0.0000
2.35	0.0954	0.2241	0.2633	0.2063	0.1212	0.0570	0.0223	0.0075	0.0022	0.0006	0.0001	0.0000	0.0000	0.0000
2.40	0.0907	0.2177	0.2613	0.2090	0.1254	0.0602	0.0241	0.0083	0.0025	0.0007	0.0002	0.0000	0.0000	0.0000
2.45	0.0863	0.2114	0.2590	0.2115	0.1295	0.0635	0.0259	0.0091	0.0028	0.0008	0.0002	0.0000	0.0000	0.0000
2.50	0.0821	0.2052	0.2565	0.2138	0.1336	0.0668	0.0278	0.0099	0.0031	0.0009	0.0002	0.0000	0.0000	0.0000
2.55	0.0781	0.1991	0.2539	0.2158	0.1376	0.0702	0.0298	0.0109	0.0035	0.0010	0.0003	0.0001	0.0000	0.0000
2.60	0.0743	0.1931	0.2510	0.2176	0.1414	0.0735	0.0319	0.0118	0.0038	0.0011	0.0003	0.0001	0.0000	0.0000
2.65	0.0707	0.1872	0.2481	0.2191	0.1452	0.0769	0.0340	0.0129	0.0043	0.0013	0.0003	0.0001	0.0000	0.0000
2.70	0.0672	0.1815	0.2450	0.2205	0.1488	0.0804	0.0362	0.0139	0.0047	0.0014	0.0004	0.0001	0.0000	0.0000
2.75	0.0639	0.1758	0.2417	0.2216	0.1523	0.0838	0.0384	0.0151	0.0052	0.0016	0.0004	0.0001	0.0000	0.0000
2.80	0.0608	0.1703	0.2384	0.2225	0.1557	0.0872	0.0407	0.0163	0.0057	0.0018	0.0005	0.0001	0.0000	0.0000
2.85	0.0578	0.1649	0.2349	0.2232	0.1590	0.0906	0.0431	0.0175	0.0062	0.0020	0.0006	0.0001	0.0000	0.0000
2.90	0.0550	0.1596	0.2314	0.2237	0.1622	0.0940	0.0455	0.0188	0.0068	0.0022	0.0006	0.0002	0.0000	0.0000
2.95	0.0523	0.1544	0.2277	0.2239	0.1652	0.0974	0.0479	0.0202	0.0074	0.0024	0.0007	0.0002	0.0000	0.0000
3.00	0.0498	0.1494	0.2240	0.2240	0.1680	0.1008	0.0504	0.0216	0.0081	0.0027	0.0008	0.0002	0.0001	0.0000
3.05	0.0474	0.1444	0.2203	0.2239	0.1708	0.1042	0.0530	0.0231	0.0088	0.0030	0.0009	0.0003	0.0001	0.0000
3.10	0.0450	0.1397	0.2165	0.2237	0.1733	0.1075	0.0555	0.0246	0.0095	0.0033	0.0010	0.0003	0.0001	0.0000
3.15	0.0429	0.1350	0.2126	0.2232	0.1758	0.1108	0.0581	0.0262	0.0103	0.0036	0.0011	0.0003	0.0001	0.0000
3.20	0.0408	0.1304	0.2087	0.2226	0.1781	0.1140	0.0608	0.0278	0.0111	0.0040	0.0013	0.0004	0.0001	0.0000
3.25	0.0388	0.1260	0.2048	0.2218	0.1802	0.1172	0.0635	0.0295	0.0120	0.0043	0.0014	0.0004	0.0001	0.0000
3.30	0.0639	0.1217	0.2008	0.2209	0.1823	0.1203	0.0662	0.0312	0.0129	0.0047	0.0016	0.0005	0.0001	0.0000
3.35	0.0351	0.1175	0.1969	0.2198	0.1841	0.1234	0.0689	0.0330	0.0138	0.0051	0.0017	0.0005	0.0001	0.0000
3.40	0.0334	0.1135	0.1929	0.2186	0.1858	0.1264	0.0716	0.0348	0.0148	0.0056	0.0019	0.0006	0.0002	0.0000
3.45	0.0317	0.1095	0.1889	0.2173	0.1874	0.1293	0.0743	0.0366	0.0158	0.0061	0.0021	0.0007	0.0002	0.0001
3.50	0.0302	0.1057	0.1850	0.2158	0.1888	0.1322	0.0771	0.0385	0.0169	0.0066	0.0023	0.0007	0.0002	0.0001

3.55	0.0001	0.0002	0.0008	0.0025	0.0071	0.0180	0.0405	0.0799	0.1350	0.1901	0.2142	0.1810	0.1020	0.0287
3.60	0.0001	0.0003	0.0009	0.0028	0.0076	0.0191	0.0425	0.0826	0.1377	0.1912	0.2125	0.1771	0.0984	0.0273
3.65	0.0001	0.0003	0.0010	0.0030	0.0082	0.0203	0.0445	0.0854	0.1403	0.1922	0.2106	0.1731	0.0949	0.0260
3.70	0.0001	0.0003	0.0011	0.0033	0.0089	0.0215	0.0466	0.0881	0.1429	0.1931	0.2087	0.1692	0.0915	0.0247
3.75	0.0001	0.0004	0.0012	0.0036	0.0095	0.0228	0.0487	0.0908	0.1453	0.1938	0.2067	0.1654	0.0882	0.0235
3.80	0.0001	0.0004	0.0013	0.0039	0.0102	0.0241	0.0508	0.0936	0.1477	0.1944	0.2046	0.1615	0.0850	0.0224
3.85	0.0001	0.0005	0.0015	0.0042	0.0109	0.0255	0.0529	0.0962	0.1500	0.1948	0.2024	0.1577	0.0819	0.0213
3.90	0.0001	0.0005	0.0016	0.0045	0.0116	0.0269	0.0551	0.0989	0.1522	0.1951	0.2001	0.1539	0.0789	0.0202
3.95	0.0002	0.0006	0.0018	0.0049	0.0124	0.0283	0.0573	0.1016	0.1543	0.1953	0.1978	0.1502	0.0761	0.0193
4.00	0.0002	0.0006	0.0019	0.0053	0.0132	0.0298	0.0595	0.1042	0.1563	0.1954	0.1954	0.1465	0.0733	0.0183
4.20	0.0002	0.0009	0.0027	0.0071	0.0168	0.0360	0.0686	0.1143	0.1633	0.1944	0.1852	0.1323	0.0630	0.0150
4.40	0.0003	0.0013	0.0037	0.0092	0.0209	0.0428	0.0778	0.1237	0.1687	0.1917	0.1743	0.1188	0.0540	0.0123
4.60	0.0005	0.0019	0.0049	0.0118	0.0255	0.0500	0.0869	0.1323	0.1725	0.1875	0.1631	0.1063	0.0462	0.0101
4.80	0.0007	0.0026	0.0064	0.0147	0.0307	0.0575	0.0959	0.1398	0.1747	0.1820	0.1517	0.0948	0.0395	0.0082
5.00	0.0009	0.0034	0.0082	0.0181	0.0363	0.0653	0.1044	0.1462	0.1755	0.1755	0.1404	0.0842	0.0337	0.0067
5.20	0.0013	0.0045	0.0104	0.0220	0.0423	0.0731	0.1125	0.1515	0.1748	0.1681	0.1293	0.0746	0.0287	0.0055
5.40	0.0018	0.0058	0.0129	0.0262	0.0486	0.0810	0.1200	0.1555	0.1728	0.1600	0.1185	0.0659	0.0244	0.0045
5.60	0.0024	0.0073	0.0157	0.0309	0.0552	0.0887	0.1267	0.1584	0.1697	0.1515	0.1082	0.0580	0.0207	0.0037
5.80	0.0032	0.0092	0.0190	0.0359	0.0620	0.0962	0.1326	0.1601	0.1656	0.1428	0.0985	0.0509	0.0176	0.0030
6.00	0.0041	0.0113	0.0225	0.0413	0.0688	0.1033	0.1377	0.1606	0.1606	0.1339	0.0892	0.0446	0.0149	0.0025
6.20	0.0052	0.0137	0.0265	0.0469	0.0757	0.1099	0.1418	0.1601	0.1549	0.1249	0.0806	0.0390	0.0126	0.0020
6.40	0.0065	0.0164	0.0307	0.0528	0.0825	0.1160	0.1450	0.1586	0.1487	0.1162	0.0726	0.0340	0.1006	0.0017
6.60	0.0081	0.0194	0.0353	0.0588	0.0891	0.1215	0.1472	0.1562	0.1420	0.1076	0.0652	0.0296	0.0090	0.0014
6.80	0.0099	0.0227	0.0401	0.0649	0.0954	0.1263	0.1486	0.1529	0.1349	0.0992	0.0584	0.0258	0.0076	0.0011
7.00	0.0119	0.0263	0.0452	0.0710	0.1014	0.1304	0.1490	0.1490	0.1277	0.0912	0.0521	0.0223	0.0064	0.0009
7.20	0.0142	0.0303	0.0504	0.0770	0.1070	0.1337	0.1486	0.1445	0.1204	0.0836	0.0464	0.0194	0.0054	0.0007
7.40	0.0168	0.0344	0.0558	0.0829	0.1121	0.1363	0.1474	0.1394	0.1130	0.0764	0.0413	0.0167	0.0045	0.0006
7.60	0.0196	0.0388	0.0613	0.0887	0.1167	0.1381	0.1454	0.1339	0.1057	0.0696	0.0366	0.0145	0.0038	0.0005
7.80	0.0227	0.0434	0.0667	0.0941	0.1207	0.1392	0.1428	0.1282	0.0986	0.0632	0.0324	0.0125	0.0032	0.0004
8.00	0.0296	0.0481	0.0722	0.0993	0.1241	0.1396	0.1396	0.1221	0.0916	0.0573	0.0286	0.0107	0.0027	0.0003

Table B   Exact Poisson Probabilities (continued)

np	14	15	16	17	18	19	20
3.10	0.0000	0.0000	0.0000	0.0000	0.0000	0.0000	0.0000
3.15	0.0000	0.0000	0.0000	0.0000	0.0000	0.0000	0.0000
3.20	0.0000	0.0000	0.0000	0.0000	0.0000	0.0000	0.0000
3.25	0.0000	0.0000	0.0000	0.0000	0.0000	0.0000	0.0000
3.30	0.0000	0.0000	0.0000	0.0000	0.0000	0.0000	0.0000
3.35	0.0000	0.0000	0.0000	0.0000	0.0000	0.0000	0.0000
3.40	0.0000	0.0000	0.0000	0.0000	0.0000	0.0000	0.0000
3.45	0.0000	0.0000	0.0000	0.0000	0.0000	0.0000	0.0000
3.50	0.0000	0.0000	0.0000	0.0000	0.0000	0.0000	0.0000
3.55	0.0000	0.0000	0.0000	0.0000	0.0000	0.0000	0.0000
3.60	0.0000	0.0000	0.0000	0.0000	0.0000	0.0000	0.0000
3.65	0.0000	0.0000	0.0000	0.0000	0.0000	0.0000	0.0000
3.70	0.0000	0.0000	0.0000	0.0000	0.0000	0.0000	0.0000
3.75	0.0000	0.0000	0.0000	0.0000	0.0000	0.0000	0.0000
3.80	0.0000	0.0000	0.0000	0.0000	0.0000	0.0000	0.0000
3.85	0.0000	0.0000	0.0000	0.0000	0.0000	0.0000	0.0000
3.90	0.0000	0.0000	0.0000	0.0000	0.0000	0.0000	0.0000
3.95	0.0000	0.0000	0.0000	0.0000	0.0000	0.0000	0.0000
4.00	0.0001	0.0000	0.0000	0.0000	0.0000	0.0000	0.0000
4.20	0.0001	0.0000	0.0000	0.0000	0.0000	0.0000	0.0000
4.40	0.0001	0.0000	0.0000	0.0000	0.0000	0.0000	0.0000
4.60	0.0002	0.0001	0.0000	0.0000	0.0000	0.0000	0.0000
4.80	0.0003	0.0001	0.0000	0.0000	0.0000	0.0000	0.0000
5.00	0.0005	0.0002	0.0000	0.0000	0.0000	0.0000	0.0000
5.20	0.0007	0.0002	0.0001	0.0000	0.0000	0.0000	0.0000

np						
5.40	0.0009	0.0003	0.0001	0.0000	0.0000	0.0000
5.60	0.0013	0.0005	0.0002	0.0001	0.0000	0.0000
5.80	0.0017	0.0007	0.0002	0.0001	0.0000	0.0000
6.00	0.0022	0.0009	0.0003	0.0001	0.0000	0.0000
6.20	0.0029	0.0012	0.0005	0.0002	0.0001	0.0000
6.40	0.0037	0.0016	0.0006	0.0002	0.0001	0.0000
6.60	0.0046	0.0020	0.0008	0.0003	0.0001	0.0000
6.80	0.0058	0.0026	0.0011	0.0004	0.0002	0.0001
7.00	0.0071	0.0033	0.0014	0.0006	0.0002	0.0001
7.20	0.0086	0.0041	0.0019	0.0008	0.0003	0.0001
7.40	0.0104	0.0051	0.0024	0.0010	0.0004	0.0002
7.60	0.0123	0.0062	0.0030	0.0013	0.0006	0.0002
7.80	0.0145	0.0075	0.0037	0.0017	0.0007	0.0003
8.00	0.0169	0.0090	0.0045	0.0021	0.0009	0.0004

								c						
np	0	1	2	3	4	5	6	7	8	9	10	11	12	13
8.20	0.0003	0.0023	0.0092	0.0252	0.0517	0.0849	0.1160	0.1358	0.1392	0.1269	0.1040	0.0776	0.0530	0.0334
8.40	0.0002	0.0019	0.0079	0.0222	0.0466	0.0784	0.1097	0.1317	0.1382	0.1290	0.1084	0.0828	0.0579	0.0374
8.60	0.0002	0.0016	0.0068	0.0195	0.0420	0.0722	0.1034	0.1271	0.1366	0.1306	0.1123	0.0878	0.0629	0.0416
8.80	0.0002	0.0013	0.0058	0.0171	0.0377	0.0663	0.0972	0.1222	0.1344	0.1315	0.1157	0.0925	0.0679	0.0459
9.00	0.0001	0.0011	0.0050	0.0150	0.0337	0.0607	0.0911	0.1171	0.1318	0.1318	0.1186	0.0970	0.0728	0.0504
9.50	0.0001	0.0007	0.0034	0.0107	0.0254	0.0483	0.0764	0.1037	0.1232	0.1300	0.1235	0.1067	0.0844	0.0617
10.00	0.0000	0.0005	0.0023	0.0076	0.0189	0.0378	0.0631	0.0901	0.1126	0.1251	0.1251	0.1137	0.0948	0.0729
10.50	0.0000	0.0003	0.0015	0.0053	0.0139	0.0293	0.0513	0.0769	0.1009	0.1177	0.1236	0.1180	0.1032	0.0834
11.00	0.0000	0.0002	0.0010	0.0037	0.0102	0.0224	0.0411	0.0646	0.0888	0.1085	0.1194	0.1194	0.1094	0.0926
11.50	0.0000	0.0001	0.0007	0.0026	0.0074	0.0170	0.0325	0.0535	0.0769	0.0982	0.1129	0.1181	0.1131	0.1001
12.00	0.0000	0.0001	0.0004	0.0018	0.0053	0.0127	0.0255	0.0437	0.0655	0.0874	0.1048	0.1144	0.1144	0.1056
12.50	0.0000	0.0000	0.0003	0.0012	0.0038	0.0095	0.0197	0.0353	0.0551	0.0765	0.0956	0.1087	0.1132	0.1089
13.00	0.0000	0.0000	0.0002	0.0008	0.0027	0.0070	0.0152	0.0281	0.0457	0.0661	0.0859	0.1015	0.1099	0.1099

Table B Exact Poisson Probabilities (continued)

np	0	1	2	3	4	5	6	7	8	9	10	11	12	13
13.50	0.0000	0.0000	0.0001	0.0006	0.0019	0.0051	0.0115	0.0222	0.0375	0.0563	0.0760	0.0932	0.1049	0.1089
14.00	0.0000	0.0000	0.0001	0.0004	0.0013	0.0037	0.0087	0.0174	0.0304	0.0473	0.0663	0.0844	0.0984	0.1060
14.50	0.0000	0.0000	0.0001	0.0003	0.0009	0.0027	0.0065	0.0135	0.0244	0.0394	0.0571	0.0753	0.0910	0.1014
15.00	0.0000	0.0000	0.0000	0.0002	0.0006	0.0019	0.0048	0.0104	0.0194	0.0324	0.0486	0.0663	0.0829	0.0956
15.50	0.0000	0.0000	0.0000	0.0001	0.0004	0.0014	0.0036	0.0079	0.0153	0.0264	0.0409	0.0577	0.0745	0.0888
16.00	0.0000	0.0000	0.0000	0.0001	0.0003	0.0010	0.0026	0.0060	0.0120	0.0213	0.0341	0.0496	0.0661	0.0814
16.50	0.0000	0.0000	0.0000	0.0001	0.0002	0.0007	0.0019	0.0045	0.0093	0.0171	0.0281	0.0422	0.0580	0.0736
17.00	0.0000	0.0000	0.0000	0.0000	0.0001	0.0005	0.0014	0.0034	0.0072	0.0135	0.0230	0.0355	0.0504	0.0658
17.50	0.0000	0.0000	0.0000	0.0000	0.0001	0.0003	0.0010	0.0025	0.0055	0.0107	0.0186	0.0297	0.0432	0.0582
18.00	0.0000	0.0000	0.0000	0.0000	0.0001	0.0002	0.0007	0.0019	0.0042	0.0083	0.0150	0.0245	0.0368	0.0509
18.50	0.0000	0.0000	0.0000	0.0000	0.0000	0.0002	0.0005	0.0014	0.0031	0.0065	0.0120	0.0201	0.0310	0.0441
19.00	0.0000	0.0000	0.0000	0.0000	0.0000	0.0001	0.0004	0.0010	0.0024	0.0050	0.0095	0.0164	0.0259	0.0378
19.50	0.0000	0.0000	0.0000	0.0000	0.0000	0.0001	0.0003	0.0007	0.0018	0.0038	0.0074	0.0132	0.0214	0.0322
20.00	0.0000	0.0000	0.0000	0.0000	0.0000	0.0001	0.0002	0.0005	0.0013	0.0029	0.0058	0.0106	0.0176	0.0271
20.50	0.0000	0.0000	0.0000	0.0000	0.0000	0.0001	0.0001	0.0004	0.0010	0.0022	0.0045	0.0084	0.0144	0.0227
21.00	0.0000	0.0000	0.0000	0.0000	0.0000	0.0000	0.0001	0.0003	0.0007	0.0017	0.0035	0.0067	0.0116	0.0188
21.50	0.0000	0.0000	0.0000	0.0000	0.0000	0.0000	0.0001	0.0002	0.0005	0.0012	0.0027	0.0052	0.0094	0.0155
22.00	0.0000	0.0000	0.0000	0.0000	0.0000	0.0000	0.0000	0.0001	0.0004	0.0009	0.0020	0.0041	0.0075	0.0127
22.50	0.0000	0.0000	0.0000	0.0000	0.0000	0.0000	0.0000	0.0001	0.0003	0.0007	0.0016	0.0032	0.0059	0.0103
23.00	0.0000	0.0000	0.0000	0.0000	0.0000	0.0000	0.0000	0.0001	0.0002	0.0005	0.0012	0.0024	0.0047	0.0083
23.50	0.0000	0.0000	0.0000	0.0000	0.0000	0.0000	0.0000	0.0001	0.0001	0.0004	0.0009	0.0019	0.0037	0.0067
24.00	0.0000	0.0000	0.0000	0.0000	0.0000	0.0000	0.0000	0.0000	0.0001	0.0003	0.0007	0.0014	0.0029	0.0053
24.50	0.0000	0.0000	0.0000	0.0000	0.0000	0.0000	0.0000	0.0000	0.0001	0.0002	0.0005	0.0011	0.0022	0.0042
25.00	0.0000	0.0000	0.0000	0.0000	0.0000	0.0000	0.0000	0.0000	0.0001	0.0001	0.0004	0.0008	0.0017	0.0033
25.50	0.0000	0.0000	0.0000	0.0000	0.0000	0.0000	0.0000	0.0000	0.0000	0.0001	0.0003	0.0006	0.0013	0.0026
26.00	0.0000	0.0000	0.0000	0.0000	0.0000	0.0000	0.0000	0.0000	0.0000	0.0001	0.0002	0.0005	0.0010	0.0020

c

np	14	15	16	17	18	19	20	21	22	23	24	25
8.20	0.0196	0.0107	0.0055	0.0026	0.0012	0.0005	0.0002	0.0001	0.0000	0.0000	0.0000	0.0000
8.40	0.0225	0.0126	0.0066	0.0033	0.0015	0.0007	0.0003	0.0001	0.0000	0.0000	0.0000	0.0000
8.60	0.0256	0.0147	0.0079	0.0040	0.0019	0.0009	0.0004	0.0002	0.0001	0.0000	0.0000	0.0000
8.80	0.0289	0.0169	0.0093	0.0048	0.0024	0.0011	0.0005	0.0002	0.0001	0.0000	0.0000	0.0000
9.00	0.0324	0.0194	0.0109	0.0058	0.0029	0.0014	0.0006	0.0003	0.0001	0.0001	0.0000	0.0000
9.50	0.0419	0.0265	0.0157	0.0088	0.0046	0.0023	0.0011	0.0005	0.0002	0.0001	0.0001	0.0000
10.00	0.0521	0.0347	0.0217	0.0128	0.0071	0.0037	0.0019	0.0009	0.0004	0.0002	0.0001	0.0000
10.50	0.0625	0.0438	0.0287	0.0177	0.0104	0.0057	0.0030	0.0015	0.0007	0.0003	0.0001	0.0001
11.00	0.0728	0.0534	0.0367	0.0237	0.0145	0.0084	0.0046	0.0024	0.0012	0.0006	0.0003	0.0001
11.50	0.0822	0.0630	0.0453	0.0306	0.0196	0.0119	0.0068	0.0037	0.0020	0.0010	0.0005	0.0002
12.00	0.0905	0.0724	0.0543	0.0383	0.0255	0.0161	0.0097	0.0055	0.0030	0.0016	0.0008	0.0004
12.50	0.0972	0.0810	0.0633	0.0465	0.0323	0.0213	0.0133	0.0079	0.0045	0.0024	0.0013	0.0006
13.00	0.1021	0.0885	0.0719	0.0550	0.0397	0.0272	0.0177	0.0109	0.0065	0.0037	0.0020	0.0010
13.50	0.1050	0.0945	0.0798	0.0633	0.0475	0.0337	0.0228	0.0146	0.0090	0.0053	0.0030	0.0016
14.00	0.1060	0.0989	0.0866	0.0713	0.0554	0.0409	0.0286	0.0191	0.0121	0.0074	0.0043	0.0024
14.50	0.1051	0.1016	0.0920	0.0785	0.0632	0.0483	0.0350	0.0242	0.0159	0.0100	0.0061	0.0035
15.00	0.1024	0.1024	0.0960	0.0847	0.0706	0.0557	0.0418	0.0299	0.0204	0.0133	0.0083	0.0050
15.50	0.0983	0.1016	0.0984	0.0897	0.0773	0.0630	0.0489	0.0361	0.0254	0.0171	0.0111	0.0069
16.00	0.0930	0.0992	0.0992	0.0934	0.0830	0.0699	0.0559	0.0426	0.0310	0.0216	0.0144	0.0092
16.50	0.0868	0.0955	0.0985	0.0956	0.0876	0.0761	0.0628	0.0493	0.0370	0.0265	0.0182	0.0120
17.00	0.0800	0.0906	0.0963	0.0963	0.0909	0.0814	0.0692	0.0560	0.0433	0.0320	0.0226	0.0154
17.50	0.0728	0.0849	0.0929	0.0956	0.0929	0.0856	0.0749	0.0624	0.0496	0.0378	0.0275	0.0193
18.00	0.0655	0.0786	0.0884	0.0936	0.0936	0.0887	0.0798	0.0684	0.0560	0.0438	0.0328	0.0237
18.50	0.0583	0.0719	0.0831	0.0904	0.0930	0.0905	0.0837	0.0738	0.0620	0.0499	0.0385	0.0285
19.00	0.0514	0.0650	0.0772	0.0863	0.0911	0.0911	0.0866	0.0783	0.0676	0.0559	0.0442	0.0336
19.50	0.0448	0.0582	0.0710	0.0814	0.0882	0.0905	0.0883	0.0820	0.0727	0.0616	0.0500	0.0390
20.00	0.0387	0.0516	0.0646	0.0760	0.0844	0.0888	0.0888	0.0846	0.0769	0.0669	0.0557	0.0446

Table B Exact Poisson Probabilities (continued)

np							c						
	14	15	16	17	18	19	20	21	22	23	24	25
22.00	0.0199	0.0292	0.0401	0.0520	0.0635	0.0735	0.0809	0.0847	0.0847	0.0810	0.0743	0.0654
22.50	0.0165	0.0248	0.0349	0.0462	0.0577	0.0684	0.0769	0.0824	0.0843	0.0824	0.0773	0.0695
23.00	0.0136	0.0209	0.0301	0.0407	0.0520	0.0629	0.0724	0.0793	0.0829	0.0829	0.0794	0.0731
23.50	0.0112	0.0175	0.0257	0.0356	0.0464	0.0574	0.0675	0.0755	0.0807	0.0824	0.0807	0.0759
24.00	0.0091	0.0146	0.0219	0.0309	0.0412	0.0520	0.0624	0.0713	0.0778	0.0812	0.0812	0.0779
24.50	0.0074	0.0120	0.0184	0.0266	0.0362	0.0466	0.0571	0.0667	0.0742	0.0791	0.0807	0.0791
25.00	0.0059	0.0099	0.0155	0.0227	0.0316	0.0415	0.0519	0.0618	0.0702	0.0763	0.0795	0.0795
25.50	0.0047	0.0081	0.0129	0.0193	0.0273	0.0367	0.0468	0.0568	0.0659	0.0730	0.0776	0.0791
26.00	0.0038	0.0066	0.0106	0.0163	0.0235	0.0322	0.0418	0.0518	0.0612	0.0692	0.0750	0.0780

Table C   Standard Normal Curve Probabilities

z	Prob	Tail	z	Prob	Tail	z	Prob	Tail
0.00	.0000	.5000	0.43	.1664	.3336	0.86	.3051	.1949
0.01	.0040	.4960	0.44	.1700	.3300	0.87	.3079	.1921
0.02	.0080	.4920	0.45	.1736	.3264	0.88	.3106	.1894
0.03	.0120	.4880	0.46	.1772	.3228	0.89	.3133	.1867
0.04	.0160	.4840	0.47	.1808	.3192	0.90	.3159	.1841
0.05	.0199	.4801	0.48	.1844	.3156	0.91	.3186	.1814
0.06	.0239	.4761	0.49	.1879	.3121	0.92	.3212	.1788
0.07	.0279	.4721	0.50	.1915	.3085	0.93	.3238	.1762
0.08	.0319	.4681	0.51	.1950	.3050	0.94	.3264	.1736
0.09	.0359	.4641	0.52	.1985	.3015	0.95	.3289	.1711
0.10	.0389	.4602	0.53	.2019	.2981	0.96	.3315	.1685
0.11	.0438	.4562	0.54	.2054	.2946	0.97	.3340	.1660
0.12	.0478	.4522	0.55	.2088	.2912	0.98	.3365	.1635
0.13	.0517	.4483	0.56	.2122	.2878	0.99	.3389	.1611
0.14	.0557	.4443	0.57	.2156	.2844	1.00	.3414	.1586
0.15	.0596	.4404	0.58	.2190	.2810	1.01	.3438	.1562
0.16	.0636	.4364	0.59	.2224	.2776	1.02	.3461	.1539
0.17	.0675	.4325	0.60	.2257	.2743	1.03	.3485	.1515
0.18	.0714	.4286	0.61	.2291	.2709	1.04	.3508	.1492
0.19	.0754	.4246	0.62	.2324	.2676	1.05	.3531	.1469
0.20	.0793	.4207	0.63	.2356	.2644	1.06	.3554	.1446
0.21	.0832	.4168	0.64	.2389	.2611	1.07	.3577	.1423
0.22	.0871	.4129	0.65	.2421	.2579	1.08	.3599	.1401
0.23	.0910	.4090	0.66	.2454	.2546	1.09	.3622	.1378
0.24	.0948	.4052	0.67	.2486	.2514	1.10	.3643	.1357
0.25	.0987	.4103	0.68	.2517	.2483	1.11	.3665	.1335
0.26	.1026	.3974	0.69	.2549	.2451	1.12	.3687	.1313
0.27	.1064	.3936	0.70	.2580	.2420	1.13	.3708	.1292
0.28	.1103	.3897	0.71	.2611	.2389	1.14	.3729	.1271
0.29	.1141	.3859	0.72	.2642	.2358	1.15	.3749	.1251
0.30	.1179	.3821	0.73	.2673	.2327	1.16	.3770	.1230
0.31	.1217	.3783	0.74	.2703	.2297	1.17	.3790	.1210
0.32	.1255	.3745	0.75	.2734	.2266	1.18	.3810	.1190
0.33	.1293	.3707	0.76	.2764	.2236	1.19	.3830	.1170
0.34	.1331	.3669	0.77	.2793	.2207	1.20	.3849	.1151
0.35	.1368	.3632	0.78	.2823	.2177	1.21	.3869	.1131
0.36	.1406	.3594	0.79	.2852	.2148	1.22	.3888	.1112
0.37	.1443	.3557	0.80	.2881	.2119	1.23	.3907	.1093
0.38	.1480	.3520	0.81	.2910	.2090	1.24	.3925	.1075
0.39	.1517	.3483	0.82	.2939	.2061	1.25	.3944	.1056
0.40	.1554	.3446	0.83	.2967	.2033	1.26	.3962	.1038
0.41	.1591	.3409	0.84	.2995	.2005	1.27	.3980	.1020
0.42	.1627	.3373	0.85	.3023	.1997	1.28	.3997	.1003

Table C   Standard Normal Curve Probabilities (continued)

z	Prob	Tail	z	Prob	Tail	z	Prob	Tail
1.29	.4015	.0985	1.71	.4564	.0436	2.13	.4834	.0166
1.30	.4032	.0968	1.72	.4573	.0427	2.14	.4838	.0162
1.31	.4049	.0951	1.73	.4582	.0418	2.15	.4842	.0158
1.32	.4066	.0934	1.74	.4591	.0409	2.16	.4846	.0154
1.33	.4083	.0917	1.75	.4599	.0401	2.17	.4850	.0150
1.34	.4099	.0901	1.76	.4608	.0392	2.18	.4854	.0146
1.35	.4115	.0885	1.77	.4616	.0384	2.19	.4857	.0143
1.36	.4131	.0869	1.78	.4625	.0375	2.20	.4861	.0139
1.37	.4147	.0853	1.79	.4633	.0367	2.21	.4864	.0136
1.38	.4162	.0838	1.80	.4641	.0359	2.22	.4868	.0132
1.39	.4177	.0823	1.81	.4648	.0352	2.23	.4871	.0129
1.40	.4193	.0807	1.82	.4656	.0344	2.24	.4874	.0126
1.41	.4207	.0793	1.83	.4664	.0336	2.25	.4878	.0122
1.42	.4222	.0778	1.84	.4671	.0329	2.26	.4881	.0119
1.43	.4236	.0764	1.85	.4678	.0322	2.27	.4884	.0116
1.44	.4251	.0749	1.86	.4686	.0314	2.28	.4887	.0113
1.45	.4265	.0735	1.87	.4693	.0307	2.29	.4890	.0110
1.46	.4279	.0721	1.88	.4699	.0301	2.30	.4893	.0107
1.47	.4292	.0708	1.89	.4706	.0294	2.31	.4895	.0105
1.48	.4306	.0694	1.90	.4713	.0287	2.32	.4898	.0102
1.49	.4319	.0681	1.91	.4719	.0281	2.33	.4901	.0099
1.50	.4332	.0668	1.92	.4726	.0274	2.34	.4903	.0097
1.51	.4345	.0655	1.93	.4732	.0268	2.35	.4906	.0094
1.52	.4357	.0643	1.94	.4738	.0262	2.36	.4909	.0091
1.53	.4370	.0630	1.95	.4744	.0256	2.37	.4911	.0089
1.54	.4382	.0618	1.96	.4750	.0250	2.38	.4913	.0087
1.55	.4394	.0606	1.97	.4756	.0244	2.39	.4916	0084
1.56	.4406	.0594	1.98	.4761	.0239	2.40	.4918	.0082
1.57	.4418	.0582	1.99	.4767	.0233	2.41	.4920	.0080
1.58	.4429	.0571	2.00	.4772	.0228	2.42	.4922	.0078
1.59	.4441	.0559	2.01	.4778	.0222	2.43	.4924	.0076
1.60	.4452	.0548	2.02	.4783	.0217	2.44	.4926	.0074
1.61	.4463	.0537	2.03	.4788	.0212	2.45	.4928	.0072
1.62	.4474	.0526	2.04	.4793	.0207	2.46	.4930	.0070
1.63	.4485	.0515	2.05	.4798	.0202	2.47	.4932	.0068
1.64	.4495	.0505	2.06	.4803	.0197	2.48	.4934	.0066
1.65	.4505	.0495	2.07	.4808	.0192	2.49	.4936	.0064
1.66	.4515	.0485	2.08	.4812	.0188	2.50	.4938	.0062
1.67	.4525	.0475	2.09	.4817	.0183	2.51	.4940	.0060
1.68	.4535	.0465	2.10	.4821	.0179	2.52	.4941	.0059
1.69	.4545	.0455	2.11	.4826	.0174	2.53	.4943	.0057
1.70	.4554	.0446	2.12	.4830	.0170	2.54	.4944	.0056

Table C   Standard Normal Curve Probabilities (continued)

z	Prob	Tail	z	Prob	Tail	z	Prob	Tail
2.55	.4946	.0054	2.77	.4972	.0028	2.99	.4986	.0014
2.56	.4948	.0052	2.78	.4973	.0027	3.00	.4986	.0014
2.57	.4949	.0051	2.79	.4974	.0026	3.01	.4987	.0013
2.58	.4951	.0049	2.80	.4974	.0026	3.02	.4987	.0013
2.59	.4952	.0048	2.81	.4975	.0025	3.03	.4988	.0012
2.60	.4953	.0047	2.82	.4976	.0024	3.04	.4988	.0012
2.61	.4955	.0045	2.83	.4977	.0023	3.05	.4989	.0011
2.62	.4956	.0044	2.84	.4977	.0023	3.06	.4989	.0011
2.63	.4957	.0043	2.85	.4978	.0022	3.07	.4989	.0011
2.64	.4958	.0042	2.86	.4979	.0021	3.08	.4990	.0010
2.65	.4960	.0040	2.87	.4979	.0021	3.09	.4990	.0010
2.66	.4961	.0039	2.88	.4980	.0020	3.10	.4990	.0010
2.67	.4962	.0038	2.89	.4981	.0019	3.11	.4991	.0009
2.68	.4963	.0037	2.90	.4981	.0019	3.12	.4991	.0009
2.69	.4964	.0036	2.91	.4982	.0018	3.13	.4991	.0009
2.70	.4965	.0035	2.92	.4982	.0018	3.14	.4992	.0008
2.71	.4966	.0034	2.93	.4983	.0017	3.15	.4992	.0008
2.72	.4767	.0033	2.94	.4984	.0016	3.16	.4992	.0008
2.73	.4968	.0032	2.95	.4984	.0016	3.17	.4992	.0008
2.74	.4969	.0031	2.96	.4985	.0015	3.18	.4993	.0007
2.75	.4970	.0030	2.97	.4985	.0015	3.19	.4993	.0007
2.76	.4971	.0029	2.98	.4986	.0014			

Note: The **prob** column is the area (probability) between the mean and the **z** value. The **tail** column is the area between the **z** value and infinity.

Table D   Exponential Distribution

x	e^{-x}	x	e^{-x}	x	e^{-x}
0.10	0.90484	3.90	0.02024	7.70	0.00045
0.20	0.81873	4.00	0.01832	7.80	0.00041
0.30	0.74082	4.10	0.01657	7.90	0.00037
0.40	0.67032	4.20	0.01500	8.00	0.00034
0.50	0.60653	4.30	0.01357	8.10	0.00030
0.60	0.54881	4.40	0.01228	8.20	0.00027
0.70	0.49659	4.50	0.01111	8.30	0.00025
0.80	0.44933	4.60	0.01005	8.40	0.00022
0.90	0.40657	4.70	0.00910	8.50	0.00020
1.00	0.36788	4.80	0.00823	8.60	0.00018
1.10	0.33287	4.90	0.00745	8.70	0.00017
1.20	0.30119	5.00	0.00674	8.80	0.00015
1.30	0.27253	5.10	0.00610	8.90	0.00014
1.40	0.24660	5.20	0.00552	9.00	0.00012
1.50	0.22313	5.30	0.00499	9.10	0.00011
1.60	0.20190	5.40	0.00452	9.20	0.00010
1.70	0.18268	5.50	0.00409	9.30	0.00009
1.80	0.16530	5.60	0.00370	9.40	0.00008
1.90	0.14957	5.70	0.00335	9.50	0.00007
2.00	0.13534	5.80	0.00303	9.60	0.00007
2.10	0.12246	5.90	0.00274	9.70	0.00006
2.20	0.11080	6.00	0.00248	9.80	0.00006
2.30	0.10026	6.10	0.00224	9.90	0.00005
2.40	0.09072	6.20	0.00203	10.00	0.00005
2.50	0.08208	6.30	0.00184	10.10	0.00004
2.60	0.07427	6.40	0.00166	10.20	0.00004
2.70	0.06721	6.50	0.00150	10.30	0.00003
2.80	0.06081	6.60	0.00136	10.40	0.00003
2.90	0.05502	6.70	0.00123	10.50	0.00003
3.00	0.04979	6.80	0.00111	10.60	0.00002
3.10	0.04505	6.90	0.00101	10.70	0.00002
3.20	0.04076	7.00	0.00091	10.80	0.00002
3.30	0.03688	7.10	0.00083	10.90	0.00002
3.40	0.03337	7.20	0.00075	11.00	0.00002
3.50	0.03020	7.30	0.00068	11.10	0.00002
3.60	0.02732	7.40	0.00061	11.20	0.00001
3.70	0.02472	7.50	0.00055	11.30	0.00001
3.80	0.02237	7.60	0.00050	11.40	0.00001

Table E Student's Distribution: Selected Probabilities

	Area (Probabilities) under the t Distribution Curve						
One Tail	.300	.200	.100	.050	.025	.010	.005
Two Tails	.600	.400	.200	.100	.050	.020	.010
Conf. Level	.40	.60	.80	.90	.95	.98	.99
df	Corresponding Values for t						
1	0.727	1.376	3.078	6.314	12.706	31.821	63.657
2	0.617	1.061	1.886	2.920	4.303	6.965	9.925
3	0.584	0.978	1.638	2.353	3.182	4.541	5.841
4	0.569	0.941	1.533	2.132	2.776	3.747	4.604
5	0.559	0.920	1.476	2.015	2.571	3.365	4.032
6	0.553	0.906	1.440	1.943	2.447	3.143	3.707
7	0.549	0.896	1.415	1.895	2.365	2.998	3.449
8	0.546	0.889	1.397	1.860	2.306	2.896	3.355
9	0.543	0.883	1.383	1.833	2.262	2.821	3.250
10	9.542	0.879	1.372	1.812	2.228	2.764	3.169
11	0.540	0.876	1.363	1.796	2.201	2.718	3.106
12	0.539	0.873	1.356	1.782	2.179	2.681	3.055
13	0.537	0.870	1.350	1.771	2.160	2.650	3.012
14	0.537	0.868	1.345	1.761	2.145	2.624	2.977
15	0.536	0.866	1.341	1.753	2.131	2.602	2.947
16	0.535	0.865	1.337	1.746	2.120	2.583	2.921
17	0.534	0.863	1.333	1.740	2.110	2.567	2.898
18	0.534	0.862	1.330	1.734	2.101	2.552	2.878
19	0.533	0.861	1.328	1.729	2.093	2.539	2.861
20	0.533	0.860	1.325	1.725	2.086	2.528	2.845
21	0.532	0.859	1.323	1.721	2.080	2.518	2.831
22	0.532	0.858	1.321	1.717	2.074	2.508	2.819
23	0.532	0.858	1.319	1.714	2.069	2.500	2.807
24	0.531	0.857	1.318	1.711	2.064	2.492	2.797
25	0.531	0.856	1.316	1.708	2.060	2.485	2.787
26	0.531	0.856	1.315	1.706	2.056	2.479	2.779
27	0.531	0.855	1.314	1.703	2.052	2.473	2.771
28	0.530	0.855	1.313	1.701	2.048	2.467	2.763
29	0.530	0.854	1.311	1.699	2.045	2.462	2.756
30	0.530	0.854	1.310	1.697	2.042	2.457	2.750
40	0.529	0.851	1.303	1.684	2.021	2.423	2.704
60	0.527	0.848	1.296	1.671	2.000	2.390	2.660
80	0.526	0.846	1.292	1.664	2.990	2.374	2.639
100	0.526	0.845	1.290	1.660	1.984	2.364	2.626
∞	0.524	0.842	1.282	1.645	1.960	2.326	2.576

Note: As the df approaches infinity, the t values become the same as the standard normal curve.

Table F   Chi-Square Distribution (Values of χ_α^2)

df	α									
	0.995	0.99	0.975	0.95	0.9	0.1	0.05	0.025	0.01	0.005
1	0.00	0.00	0.00	0.00	0.02	2.71	3.84	5.02	6.63	7.88
2	0.01	0.02	0.05	0.10	0.21	4.61	5.99	7.38	9.21	10.60
3	0.07	0.11	0.22	0.35	0.58	6.25	7.81	9.35	11.34	12.84
4	0.21	0.30	0.48	0.71	1.06	7.78	9.49	11.14	13.28	14.86
5	0.41	0.55	0.83	1.15	1.61	9.24	11.07	12.83	15.09	16.75
6	0.68	0.87	1.24	1.64	2.20	10.64	12.59	14.45	16.81	18.55
7	0.99	1.24	1.69	2.17	2.83	12.02	14.07	16.01	18.48	20.28
8	1.34	1.65	2.18	2.73	3.49	13.36	15.51	17.54	20.01	21.96
9	1.73	2.09	2.70	3.33	4.17	14.68	16.92	19.02	21.67	23.59
10	2.16	2.56	3.25	3.94	4.87	15.99	18.31	20.48	23.21	25.19
11	2.60	3.05	3.82	4.57	5.58	17.28	19.68	21.92	24.72	26.76
12	3.07	3.57	4.40	5.23	6.30	18.55	21.03	23.34	26.22	28.30
13	3.57	4.11	5.01	5.89	7.04	19.81	22.36	24.74	27.69	29.82
14	4.07	4.66	5.63	6.57	7.79	21.06	23.68	26.12	29.14	31.32
15	4.60	5.23	6.26	7.26	8.55	22.31	25.00	27.49	30.58	32.80
16	5.14	5.81	6.91	7.96	9.31	23.54	26.30	28.85	32.00	34.27

17	5.70	6.41	7.56	8.67	10.09	24.77	27.59	30.19	33.41	35.72
18	6.26	7.01	8.23	9.39	10.86	25.99	28.87	31.53	34.81	37.16
19	6.84	7.63	8.91	10.12	11.65	27.20	30.14	32.85	36.19	38.58
20	7.43	8.26	9.59	10.85	12.44	28.41	31.41	34.17	37.57	40.00
21	8.03	8.90	10.28	11.59	13.24	29.62	32.67	35.48	38.93	41.40
22	8.64	9.54	10.98	12.34	14.04	30.81	33.92	36.78	40.29	42.80
23	9.26	10.20	11.69	13.09	14.85	32.01	35.17	38.08	41.64	44.18
24	9.89	10.76	12.40	13.85	15.66	33.20	36.42	39.36	42.98	45.56
25	10.52	11.52	13.12	14.61	16.47	34.38	37.65	40.65	44.31	46.93
26	11.16	12.20	13.84	15.38	17.29	35.56	38.89	41.92	45.64	48.29
27	11.81	12.88	14.57	16.15	18.11	36.74	40.11	43.19	46.96	49.65
28	12.46	13.56	15.31	16.93	18.94	37.92	41.34	44.46	48.28	50.99
29	13.12	14.26	16.05	17.71	19.77	39.09	42.56	45.72	49.59	52.34
30	13.79	14.95	16.79	18.49	20.30	40.26	43.77	46.98	50.89	53.67
50	27.99	29.71	32.36	34.76	37.69	63.17	67.50	71.42	76.15	79.49
100	67.33	70.06	74.22	77.93	82.36	118.50	124.30	129.60	135.80	140.20
500	422.30	429.40	439.90	449.10	459.90	540.90	553.10	563.90	576.50	585.20
1000	888.60	898.80	914.30	927.60	943.10	1058.00	1075.00	1090.00	1107.00	1119.00

Table G　Values of F

$F_{.01}$

df for Denom	df for Numerator								
	1	2	3	4	5	6	7	8	9
1	4052.00	4995.50	5403.00	5625.00	5764.00	5859.00	5928.00	5981.00	6022.00
2	98.50	99.00	99.17	99.25	99.30	99.33	99.36	99.37	99.39
3	34.12	30.82	29.46	28.71	28.24	27.91	27.67	27.49	27.35
4	21.20	18.00	16.69	15.98	15,52	15.21	14.98	14.80	14.66
5	16.26	13.27	12.06	11.39	10.91	10.67	10.46	10.29	10.16
6	13.75	10.92	9.78	9.15	8.75	8.47	8.26	8.10	7.98
7	12.25	9.55	8.45	7.85	7.46	7.19	6.99	6.84	6.72
8	11.26	8.65	7.59	7.01	6.63	6.37	6.18	6.03	5.91
9	10.56	8.02	6.99	6.42	6.60	5.80	5.61	5.47	5.35
10	10.04	7.56	6.55	5.99	5.64	5.39	5.20	5.06	4.94
11	9.65	7.21	6.22	5.67	5.32	5.07	4.89	4.74	4.63
12	9.33	6.93	5.95	5.41	6.06	4.82	4.64	4.50	4.39
13	9.07	6.70	5.74	5.21	4.86	4.62	4.44	4.30	4.19
14	8.86	6.51	5.56	5.04	4.69	4.46	4.28	4.14	4.03
15	8.68	6.36	5.42	4.89	4.56	4.32	4.14	4.00	3.89
16	8.53	6.23	5.29	4.77	4.44	4.20	4.03	3.89	3.78
17	8.50	6.11	5.18	4.67	4.34	4.10	3.93	3.79	3.67
18	8.29	6.01	5.09	4.58	4.25	4.01	3.84	3.71	3.60
19	8.18	5.93	5.01	4.50	4.17	3.94	3.77	3.63	5.62
20	8.10	5.85	4.94	4.43	4.10	3.87	3.70	3.56	3.46
21	8.02	5.78	4.87	4.37	4.04	3.81	3.64	3.51	3.40
22	7.95	5.72	4.82	4.31	3.99	3.76	3.59	3.45	3.35
23	7.88	5.66	4.76	4.26	3.94	3.71	3.54	3.41	3.30
24	7.82	5.61	4.72	4.22	3.90	3.67	3.50	3.36	3.26
25	7.77	5.57	4.68	4.18	3.85	3.63	3.46	3.32	3.22
26	7.72	5.53	4.64	4.14	3.82	3.59	3.42	3.29	3.18
27	7.68	5.49	4.60	4.11	3.78	3.56	3.39	3.26	3.15
28	7.64	5.45	4.57	4.07	3.75	3.53	3.36	3.23	3.12
29	7.60	5.42	4.54	4.04	3.73	3.50	3.33	3.20	3.10
30	7.56	5.39	4.51	4.02	3.70	3.47	3.30	3.17	3.07
40	7.31	5.18	4.31	3.83	3.51	3.29	3.12	2.99	2.89
60	7.08	4.98	4.13	3.65	3.34	3.12	2.95	2.82	2.72
120	6.85	4.79	3.95	3.48	3.17	2.96	2.79	2.66	2.56

Table G   Values of F (continued)

$F_{.01}$

df for	df for Numerator								
Denom	10	12	15	20	24	30	40	60	120
1	6056.00	6106.00	6157.00	6209.00	6235.00	6261.00	6287.00	6313.00	6339.00
2	99.40	99.42	99.43	99.45	99.46	99.47	99.47	99.48	99.49
3	27.23	27.05	26.87	26.69	26.60	26.50	2641	26.32	26.22
4	14.55	14.37	14.20	14.02	13.93	13.84	13.75	13.65	13.56
5	10.05	9.89	9.72	9.55	9.47	9.38	9.29	9.20	9.11
6	7.87	7.72	7.56	7.40	7.31	7.23	7.14	7.06	6.97
7	6.62	6.47	6.31	6.46	6.07	5.99	5.91	5.82	5.74
8	5.81	5.67	5.52	5.36	5.28	5.20	5.12	5.03	4.95
9	5.26	5.11	4.96	4.81	4.73	4.65	4.57	4.48	4.40
10	4.85	4.71	4.56	4.41	4,.33	4.25	4.17	4.08	4.00
11	4.54	4.40	4.25	4.10	4.02	3.94	3.86	3.78	3.69
12	4.30	4.16	4.01	3.86	3.78	3.70	3.62	3.54	3.45
13	4.10	3.96	3.82	3.66	3.59	3.51	3.43	3.34	3.25
14	3.94	3.80	3.66	3.51	3.43	3.35	3.27	3.18	3.09
15	3.80	3.67	3.52	3.37	3.29	3.21	3.13	3.05	2.96
16	3.69	3.55	3.41	3.26	3.18	3.10	3.02	2.93	2.84
17	3.59	3.46	3.31	3.16	3.08	3.00	2.92	2.83	2.75
18	3.51	3.37	3.23	3.08	3.00	2.92	2.84	2.75	2.66
19	3.43	3.30	3.15	3.00	2.92	2.84	2.76	2.667	2.58
20	3.37	3.23	3.09	2.94	2.86	2.78	2.69	2.61	2.52
21	3.31	3.17	3.03	2.88	2.80	2.72	2.64	2.55	2.46
22	3.26	3.12	2.98	2.83	2.75	2.67	2.58	2.50	2.40
23	3.21	3.07	2.93	2.78	2.70	2.62	2.54	2.45	2.35
24	3.17	3.03	2.89	2.74	2.66	2.58	2.49	2.40	2.31
25	3.13	2.99	2.85	2.70	2.62	2.54	2.45	2.36	2.27
26	3.09	2.96	2.81	2.66	2.58	2.50	2.42	2.33	2.23
27	3.60	2.93	2.78	2.63	2.55	2.47	2.38	2.29	2.20
28	3.03	2.90	2.75	2.60	2.52	2.44	2.35	2.26	2.17
29	3.00	2.87	2.73	2.57	2.49	2.41	2.33	2.23	2.14
30	2.98	2.84	2.70	2.55	2.47	2.39	2.30	2.21	2.11
40	2.80	2.66	2.52	2.37	2.29	2.20	2.11	2.20	1.92
60	2.63	2.50	2.35	2.20	2.12	2.03	1.94	1.84	1.73
120	2.47	2.34	2.19	2.03	1.95	1.86	1.76	1.66	1.53

Table G   Values of F

$F_{.025}$

df for	df for Numerator								
Denom	1	2	3	4	5	6	7	8	9
1	647.79	799.50	846.16	899.58	921.85	937.11	948.22	956.66	963.28
2	38.51	39.00	39.17	39.25	39.30	39.33	39.36	39.37	39.39
3	17.44	16.04	15.44	15.10	14.89	14.74	14.62	14.54	14.47
4	12.22	10.65	9.98	9.60	9.36	9.20	9.07	8.98	8.90
5	10.00	8.43	7.76	7.39	7.15	6.98	6.85	6.76	6.68
6	8.81	7.26	6.60	6.23	5.99	5.82	5.70	5.60	5.52
7	8.07	6.54	5.89	5.52	5.29	5.12	4.99	4.90	4.82
8	7.57	6.06	5.42	5.05	4.82	4.65	4.53	4.43	4.36
9	7.21	5.71	5.08	4.72	4.48	4.32	4.20	4.10	4.03
10	6.94	5.43	4.83	4.47	4.24	4.07	3.95	3.85	3.78
11	6.72	5.26	4.63	4.28	4.04	3.88	3.76	3.66	3.59
12	6.55	5.10	4.47	4.12	3.89	3.73	3.61	3.81	3.44
13	6.41	4.97	4.35	4.00	3.77	3.60	3.48	0.39	3.31
14	6.30	4.86	4.24	3.89	3.66	3.50	3.38	3.29	3.21
15	6.20	4.77	4.15	3.80	3.58	3.41	3.29	3.20	3.12
16	6.12	4.69	4.08	3.73	3.50	3.34	3.22	3.12	3.05
17	6.04	4.62	4.01	3.66	3.44	3.28	3.16	3.06	2.98
18	5.98	4.56	3.95	3.61	3.38	3.22	3.10	3.01	2.93
19	5.92	4.51	3.90	3.56	3.33	3.17	3.05	2.96	2.88
20	5.87	4.46	3.86	3.51	3.29	3.13	3.01	2.94	2.84
21	5.83	4.42	3.82	3.48	3.28	3.09	2.97	2.87	2.80
22	5.79	4.38	3.78	3.44	3.22	3.05	2.93	2.84	2.76
23	5.75	4.35	3.75	3.41	3.18	3.02	2.90	2.81	2.73
24	5.72	4.32	3.72	3.38	3.15	2.99	2.87	2.78	2.70
25	5.69	4.29	3.69	3.35	3.13	2.97	2.85	2.75	2.68
26	5.66	4.27	3.67	3.33	3.10	2.94	2.82	2.73	2.65
27	5.63	4.24	3.65	3.31	3.08	2.92	2.80	2.71	2.63
28	5.61	4.22	3.63	3.29	3.06	2.90	2.78	2.69	2.61
29	5.59	4.20	3.61	3.27	3.04	2.88	2.76	2.67	2.59
30	5.57	4.18	3.59	3.25	3.03	2.87	2.75	2.65	2.57
40	5.42	4.05	3.46	3.13	2.90	2.74	2.62	2.53	2.45
60	5.29	3.93	3.34	3.01	2.79	2.63	2.51	2.41	2.33
120	5.15	3.80	3.23	2.89	2.67	2.52	2.39	2.30	2.22

Table G Values of F (continued)

$F_{.025}$

df for	df for Numerator								
Denom	10	12	15	20	24	30	40	60	120
1	968.63	976.71	948.87	993.10	997.25	1001.40	1005.60	1009.80	1014.00
2	39.40	39.42	39.43	39.45	39.46	39.47	39.47	39.48	39.49
3	14.42	14.64	14.25	14.17	14.12	14.08	14.04	13.99	13.95
4	8.84	8.75	8.66	8.56	8.51	8.46	8.41	8.36	8.31
5	6.62	6.52	6.43	6.33	6.28	6.23	6.18	6.12	6.07
6	5.46	5.37	5.27	5.17	5.12	5.07	5.01	4.96	4.90
7	4.76	4.67	4.57	4.47	4.42	4.36	4.31	4.25	4.20
8	4.30	4.20	4.10	4.00	3.95	3.89	3.84	3.78	3.73
9	3.96	3.87	3.77	3.67	3.61	3.56	3.51	3.45	3.39
10	3.72	3.62	3.52	3.42	3.37	3.31	3.26	3.20	3.14
11	3.53	3.43	3.33	3.23	3.17	3.12	3.06	3.00	2.94
12	3.37	3.28	3.18	3.07	3.02	2.96	2.91	2.85	2.79
13	3.25	3.15	3.05	2.95	2.89	2.84	2.78	2.72	2.66
14	3.15	3.05	2.95	2.84	2.79	2.73	2.67	2.61	2.55
15	3.06	2.96	2.86	2.76	2.70	2.64	2.59	2.52	2.46
16	2.99	2.89	2.79	2.68	2.63	2.57	2.51	2.45	2.38
17	2.92	2.82	2.72	2.62	2.56	2.50	2.44	2.38	2.32
18	2.87	2.77	2.67	2.56	2.50	2.44	2.38	2.32	2.26
19	2.82	2.72	2.62	2.51	2.45	2.39	2.33	2.27	2.20
20	2.77	2.68	2.57	2.46	2.41	2.35	2.29	2.22	2.16
21	2.73	2.64	2.53	2.42	2.37	2.31	2.25	2.18	2.11
22	2.70	2.60	2.50	2.39	2.33	2.27	2.21	2.14	2.08
23	2.67	2.57	2.47	2.36	2.30	2.24	2.18	2.11	2.04
24	2.64	2.54	2.44	2.33	2.27	2.21	2.15	2.08	2.01
25	2.61	2.51	2.41	2.30	2.24	2.18	2.12	2.05	1.98
26	2.59	2.49	2.39	2.28	2.22	2.16	2.09	2.03	1.95
27	2.57	2.47	2.36	2.25	2.19	2.13	2.07	2.00	1.93
28	2.55	2.45	2.34	2.23	2.17	2.11	2.05	1.98	1.91
29	2.53	2.43	2.32	2.21	2.15	2.09	2.03	1.96	1.89
30	2.51	2.41	2.31	2.20	2.14	2.07	2.01	1.94	1.87
40	2.39	2.29	2.18	2.07	2.01	1.94	1.88	1.80	1.72
60	2.27	2.17	2.06	1.94	1.88	1.82	1.74	1.67	1.58
120	2.16	2.05	1.95	1.82	1.76	1.69	1.61	1.53	1.43

Table G Values of F

$F_{.10}$

df for Denom	df for Numerator								
	1	2	3	4	5	6	7	8	9
1	161.40	199.50	215.70	224.60	230.20	234.00	236.80	238.90	240.50
2	18.51	19.00	19.16	19.25	19.30	19.33	19.35	19.37	19.38
3	10.13	9.55	9.28	9.12	9.01	8.94	8.89	8.85	8.81
4	7.71	6.94	6.59	6.39	6.26	6.16	6.09	6.04	6.00
5	6.61	5.79	5.41	5.19	5.05	4.95	4.88	4.82	4.77
6	5.99	5.14	4.76	4.53	4.39	4.28	4.21	4.15	4.10
7	5.59	4.74	4.35	4.12	3.97	3.87	3.79	3.73	3.68
8	5.32	4.46	4.07	3.84	3.69	3.58	3.50	3.44	3.39
9	5.12	4.26	3.86	3.63	3.48	3.37	3.29	3.23	3.18
10	4.96	4.10	3.71	3.48	3.33	3.22	3.14	3.07	3.02
11	4.84	3.98	3.59	3.36	3.20	3.09	3.01	2.95	2.90
12	4.75	3.89	3.49	3.26	3.11	3.00	2.91	2.85	2.80
13	4.67	3.81	3.47	3.18	3.03	2.92	2.83	2.77	2.71
14	4.60	3.74	3.34	3.11	2.96	2.85	2.76	2.70	2.65
15	4.54	3.68	3.29	3.06	2.90	2.79	2.71	2.64	2.59
16	4.49	3.63	3.24	3.01	2.85	2.74	2.66	2.59	2.54
17	4.45	3.59	3.20	2.96	2.81	2.70	2.61	2.55	2.49
18	4.41	3.55	3.16	2.93	2.77	2.66	2.58	2.51	2.46
19	4.38	3,52	3.13	2.90	2.74	2.63	2.54	2.48	2.42
20	4.35	3.49	3.10	2.87	2.71	2.60	2.51	2.45	2.39
21	4.32	3.47	3.07	2.84	2.68	2.57	2.49	2.42	2.37
22	4.30	3.44	3.05	2.82	2.66	2.55	2.46	2.40	2.34
23	4.28	3.20	3.03	2.80	2.64	2.53	2.44	2.37	2.32
24	4.26	3.40	3.01	2.78	2.62	2.51	2.42	2.36	2.30
25	4.24	3.39	2.99	2.76	2.60	2.49	2.40	2.34	2.28
26	4.23	3.37	2.98	2.74	2.59	2.47	2.39	2.32	2.27
27	4.21	3.35	2.96	2.73	2.57	2.46	2.67	2.31	2.25
28	4.20	3.35	2.95	2.71	2.56	2.45	2.37	2.29	2.24
29	4.18	3.33	2.93	2.70	2.55	2.43	2.35	2.28	2.22
30	4.17	3.32	2.92	2.69	2.53	2.42	2.33	2.27	2.21
40	4.01	3.23	2.84	2.61	2.45	2.34	2.25	2.18	2.12
60	4.00	3.15	2.76	2.53	2.37	2.25	2.17	2.10	2.04
120	3.92	3.07	2.68	2.45	2.29	2.17	2.09	2.02	1.96

Table G   Values of F (continued)

$F_{.10}$

df for Denom	df for Numerator								
	10	12	15	20	24	30	40	60	120
1	241.90	243.90	245.90	248.00	249.10	250.10	251.10	252.20	253.30
2	19.40	19.41	19.43	19.45	19.45	19.46	19.47	19.48	19.49
3	8.79	8.74	8.70	8.66	8.64	8.62	8.69	8.57	8.55
4	5.96	5.91	5.86	5.80	5.77	5.75	5.72	5.69	5.66
5	4.74	4.68	4.62	4.56	4.53	4.50	4.46	4.43	4.40
6	4.06	4.00	3.94	3.87	3.84	3.81	3.77	3.74	3.70
7	3.64	3.57	3.51	3.41	3.41	3.38	3.34	3.30	3.27
8	3.35	3.28	3.22	3.15	3.12	3.08	3.04	3.01	2.97
9	3.14	3.07	3.01	2.94	2.90	2.86	2.83	2.79	2.75
10	2.98	2.91	2.85	2.77	2.74	2.70	2.66	2.62	2.58
11	2.85	2.79	2.72	2.65	2.61	2.57	2.53	2.49	2.45
12	2.75	2.69	2.62	2.54	2.51	2.47	2.43	2.38	2.34
13	2.67	2.60	2.53	2.46	2.42	2.38	2.34	2.30	2.25
14	2.60	2.53	2.46	2.39	2.35	2.31	2.27	2.22	2.18
15	2.54	2.48	2.40	2.33	2.29	2.25	2.20	2.16	2.11
16	2.49	2.42	2.35	2.28	2.24	2.19	2.16	2.11	2.06
17	2.45	2.38	2.31	2.23	2.19	2.15	2.10	2.06	2.01
18	2.41	2.34	2.27	2.19	2.18	2.11	2.06	2.02	1.97
19	2.38	2.31	2.23	2.16	2.11	2.07	2.03	1.98	1.93
20	2.35	2.28	2.20	2.12	2.01	2.04	1.99	1.95	1.90
21	2.32	2.25	2.12	2.10	2.05	2.01	1.96	1.92	1.87
22	2.30	2.23	2.15	2.07	2.03	1.98	1.94	1.89	1.84
23	2.27	2.20	2.13	2.05	2.01	1.96	1.91	1.86	1.81
24	2.25	2.18	2.11	2.03	1.98	1.94	1.89	1.84	1.79
25	2.24	2.16	2.09	2.01	1.96	1.92	1.87	1.82	1.77
26	2.22	2.15	2.07	1.99	1.95	1.90	1.85	1.80	1.75
27	2.20	2.13	2.06	1.97	1.93	1.88	1.84	1.79	1.73
28	2.19	2.12	2.04	1.96	1.91	1.87	1.82	1.77	1.71
29	2.18	2.10	2.03	1.94	1.90	1.85	1.81	1.75	1.70
30	2.16	2.09	2.01	1.93	1.89	1.84	1.79	1.74	1.68
40	2.08	2.00	1.92	1.84	1.79	1.74	1.69	1.64	1.58
60	1.99	1.92	1.84	1.75	1.70	1.65	1.59	1.53	1.47
120	1.91	1.83	1.75	1.66	1.61	1.55	1.50	1.43	1.35

Table H Shewhart Control Chart Constant Table

Sample Size (n)	Average and Range			Average and Standard Deviation			Std. Dev. Estimate (d_2)
	A_2	D_3	D_4	A_3	B_3	B_4	
2	1.88	0	3.27	2.66	0	3.27	1.13
3	1.02	0	2.57	1.95	0	2.57	1.69
4	0.73	0	2.28	1.63	0	2.27	2.06
5	0.58	0	2.11	1.43	0	2.09	2.33
6	0.48	0	2.00	1.29	0.03	1.97	2.53
7	0.42	0.08	1.92	1.18	0.12	1.88	2.70
8	0.37	0.14	1.86	1.10	0.19	1.82	2.85
9	0.34	0.18	1.82	1.03	0.24	1.76	2.97
10	0.31	0.22	1.78	0.98	0.28	1.72	3.08
11	0.29	0.26	1.74	0.93	0.32	1.68	3.17
12	0.27	0.28	1.72	0.89	0.35	1.65	3.26
13	0.25	0.31	1.69	0.85	0.38	1.62	3.34
14	0.24	0.33	1.67	0.82	0.41	1.59	3.41
15	0.22	0.35	1.65	0.79	0.43	1.57	3.47
16	0.21	0.36	1.64	0.76	0.45	1.55	3.53
17	0.20	0.38	1.62	0.74	0.47	1.53	3.59
18	0.19	0.39	1.61	0.72	0.48	1.52	3.64
19	0.19	0.40	1.60	0.70	0.50	1.50	3.69
20	0.18	0.41	1.59	0.68	0.51	1.49	3.74

Note: Estimate of the standard deviation using the range $s = \bar{R}/d_2$.

Table I Poisson Distribution: Selected Probabilities
(probability of c or less for specified np values)

np	0	1	2	3	4	5	6	7	8
0.02	.980	1.000							
0.04	.961	.999	1.000						
0.06	.942	.998	1.000						
0.08	.923	.997	1.000						
0.10	.905	.995	1.000						
0.15	.861	.990	.999	1.000					
0.20	.819	.982	.999	1.000					
0.25	.779	.974	.998	1.000					
9.30	.741	.963	.996	1.000					
0.35	.705	.951	.994	1.000					
0.40	.670	.938	.992	.999	1.000				
0.45	.638	.925	.989	.999	1.000				
0.50	.607	.910	.986	.998	1.000				
0.55	.577	.894	.982	.998	1.000				
0.60	.549	.878	.977	.997	1.000				
0.65	.522	.861	.972	.996	.999	1.000			
0.70	.497	.844	.966	.994	.999	1.000			
0.75	.472	.827	.959	.993	.000	1.000			
0.80	.449	.809	.953	.991	.999	1.000			
0.85	.427	.791	.945	.989	.998	1.000			
0.90	.407	.772	.937	.987	.998	1.000			
0.95	.397	.754	.929	.984	.997	1.000			
1.00	.368	.736	.920	.981	.996	.999	1.000		
1.10	.333	.699	.900	.974	.995	.999	1.000		
1.20	.301	.663	.879	.966	.993	.998	1.000		
1.30	.273	.627	.857	.957	.989	.998	1.000		
1.40	.247	.592	.833	.946	.986	.997	.999	1.000	
1.50	.223	.558	.809	.934	.981	.996	.999	1.000	
1.60	.202	.525	.783	.921	.976	.994	.999	1.000	
1.70	.183	.493	.757	.907	.970	.992	.998	1.000	
1.80	.165	.463	.731	.891	.964	.990	.997	.999	1.000
1.90	.150	.434	.704	.875	.956	.987	.997	.999	1.000
2.00	.135	.406	.677	.857	.947	.983	.995	.999	1.000
2.1	.123	.380	.650	.839	.938	.980	.995	.999	1.000
2.2	.111	.355	.623	.819	.928	.975	.993	.998	1.000
2.3	.100	.331	.596	.799	.916	.970	.991	.998	1.000

Table I   Poisson Distribution: Selected Probabilities (continued)

np	0	1	2	3	4	5	6	7	8
2.4	.091	.308	.570	.779	.904	.964	.988	.997	.999
2.5	.082	.287	.543	.757	.891	.958	.986	.996	.999
2.6	.074	.267	.518	.736	.877	.951	.983	.995	.999
2.7	.067	.249	.494	.715	.864	.944	.980	.994	.999
2.8	.061	.231	.469	.692	.848	.935	.976	.992	.998
2.9	.055	.215	.446	.670	.832	.926	.971	.990	.997
3.0	.050	.199	.423	.647	.815	.916	.966	.988	.996
3.2	.041	.171	.380	.603	.781	.895	.955	.983	.994
3.4	.033	.147	.340	.558	.744	.871	.942	.977	.992
3.6	.027	.126	.303	.515	.706	.844	.927	.969	.988
3.8	.022	.107	.269	.473	.668	.816	.909	.960	.984
4.0	.018	.092	.238	.433	.629	.785	.889	.949	.979
4.2	.015	.078	.210	.395	.590	.753	.867	.936	.972
4.4	.012	.066	.185	.359	.551	.720	.844	.921	.964
4.6	.010	.056	.163	.326	.513	.686	.818	.905	.955
4.8	.008	.048	.143	.294	.476	.651	.791	.887	.944
5.0	.007	.040	.125	.265	.440	.616	.762	.867	.932
5.2	.006	.034	.109	.238	.406	.581	.732	.845	.918
5.4	.005	.029	.095	.213	.373	.546	.702	.822	.903
5.6	.004	.024	.082	.191	.342	.512	.670	.797	.886
5.8	.003	.021	.072	.170	.313	.478	.638	.771	.867
6.0	.002	.017	.062	.151	.285	.446	.606	.744	.847
6.2	.002	.015	.054	.134	.259	.414	.574	.716	.826
6.4	.002	.012	.046	.119	.235	.384	.542	.687	.803
6.6	.001	.010	.040	.105	.213	.355	.511	.658	.780
6.8	.001	.009	.034	.093	.192	.327	.480	.628	.755
7.0	.001	.007	.030	.082	.173	.301	.450	.599	.729
7.2	.001	.006	.025	.072	.156	.276	.420	.569	.703
7.4	.001	.005	.022	.063	.140	.253	.392	.539	.676
7.6	.001	.004	.019	.055	.125	.231	.365	.510	.648
7.8	.000	.004	.016	.048	.112	.210	.338	.481	.620
8.0	.000	.003	.014	.042	.100	.191	.313	.453	.593
8.5	.000	.002	.009	.030	.074	.150	.256	.386	.523
9.0	.000	.001	.006	.021	.055	.116	.207	.324	.456
9.5	.000	.001	.004	.015	.040	.089	.165	.269	.392

Table I Poisson Distribution: Selected Probabilities (continued)

np	9	10	11	12	13	14	15	16	17
2.4	1.000								
2.5	1.000								
2.6	1.000								
2.7	1.000								
2.8	.999	1.000							
2.9	.999	1.000							
3.0	.999	1.000							
3.2	.998	1.000							
3.4	.997	.999	1.000						
3.6	.996	.999	1.000						
3.8	.994	.998	.999	1.000					
4.0	.992	.997	.999	1.000					
4.2	.989	.996	.999	1.000					
4.4	.985	.994	.998	.999	1.000				
4.6	.980	.992	.997	.999	1.000				
4.8	.975	.990	.996	.999	1.000				
5.0	.968	.986	.995	.998	.999	1.000			
5.2	.960	.982	.993	.997	.999	1.000			
5.4	.951	.977	.990	.996	.999	1.000			
5.6	.941	.972	.988	.995	.998	.999	1.000		
5.8	.929	.965	.984	.993	.997	.999	1.000		
6.0	.916	.957	.980	.991	.996	.999	.999	1.000	
6.2	.902	.949	.975	.989	.995	.998	.999	1.000	
6.4	.886	.939	.969	.986	.994	.997	.999	1.000	
6.6	.869	.927	.963	.982	.992	.997	.999	.999	1.000
6.8	.850	.915	.955	.978	.990	.996	.998	.999	1.000
7.0	.830	.901	.947	.973	.987	.994	.998	.999	1.000
7.2	.810	.887	.937	.967	.984	.993	.997	.999	.999
7.4	.788	.871	.926	.961	.980	.991	.996	.998	.999
7.6	.765	.854	.915	.954	.976	.989	.995	.998	.999
7.8	.741	.835	.902	.945	.971	.986	.993	.997	.999
8.0	.717	.816	.888	.936	.966	.983	.992	.996	.998
8.5	.653	.763	.849	.909	.949	.973	.986	.993	.997
9.0	.587	.706	.803	.876	.926	.959	.978	.989	.995
9.5	.522	.645	.752	.836	.898	.940	.967	.982	.991

Table I Poisson Distribution: Selected Probabilities (continued)

	c								
np	*0*	*1*	*2*	*3*	*4*	*5*	*6*	*7*	*8*
10.0	.000	.000	.003	.010	.029	.067	.130	.220	.333
10.5	.000	.000	.002	.007	.021	.050	.102	.179	.279
11.0	.000	.000	.001	.005	.015	.038	.079	.143	.232
11.5	.000	.000	.001	.003	.011	.028	.060	.114	.191
12.0	.000	.000	.001	.002	.008	.020	.046	.090	.155
13.0	.000	.000	.000	.001	.004	.011	.026	.054	.100
14.0	.000	.000	.000	.000	.002	.006	.014	.032	.062
15.0	.000	.000	.000	.000	.001	.003	.008	.018	.037
16.0					.000	.001	.004	.010	.022
17.0					.000	.001	.002	.005	.013
18.0					.000	.000	.001	.003	.007
19.0					.000	.000	.001	.002	.004
20.0					.000	.000	.000	.001	.002
21.0					.000	.000	.000	.000	.001
22.0					.000	.000	.000	.000	.001
23.0					.000	.000	.000	.000	.000
24.0					.000	.000	.000	.000	.000
25.0					.000	.000	.000	.000	.000

Table I   Poisson Distribution: Selected Probabilities (continued)

np	9	10	11	12	c 13	14	15	16	17
10.0	.458	.583	.697	.792	.864	.917	.951	.973	.986
10.5	.397	.521	.639	.742	.825	.888	.932	.960	.978
11.0	.341	.460	.579	.689	.781	.854	.907	.944	.968
11.5	.289	.402	.520	.633	.733	.815	.878	.924	.954
12.0	.242	.347	.462	.576	.682	.772	.844	.899	.937
13.0	.166	.252	.353	.463	.573	.675	.764	.835	.890
14.0	.109	.176	.260	.358	.464	.570	.669	.756	.827
15.0	.070	.118	.185	.268	.363	.466	.568	.664	.749
16.0	.043	.077	.127	.193	.275	.368	.467	. 566	.659
17.0	.026	.049	.085	.135	.201	.281	.371	.468	.564
18.0	.015	.030	.055	.092	.143	.208	.287	.375	.469
19.0	.009	.018	.035	.061	.098	.150	.215	.292	.378
20.0	.005	.011	.021	.039	.066	.105	.157	.221	.297
21.0	.003	.006	.013	.025	.043	.072	.111	.163	.227
22.0	.002	.004	.008	.015	.028	.048	.077	.117	.169
23.0	.001	.002	.004	.009	.017	.031	.052	.082	.123
24.0	.000	.001	.003	.005	.011	.020	.034	.056	.087
25.0	.000	.001	.001	.003	.006	.012	.022	.038	.060

Table I Poisson Distribution: Selected Probabilities (continued)

					c				
np	18	19	20	21	22	23	24	25	26
7.0									
7.2	1.000								
7.4	1.000								
7.6	1.000								
7.8	1.000								
8.0	.999	1.000	1.000						
8.5	.999	.999	1.000						
9.0	.998	.999	1.000						
9.5	.996	.998	.999	1.000					
10.0	.993	.997	.998	.999	1.000				
10.5	.988	.994	.997	.999	.999	1.000			
11.0	.982	.991	.995	.998	.999	1.000			
11.5	.974	.986	.992	.996	.998	.999	1.000		
12.0	.963	.979	.988	.994	.997	.999	.999	1.000	
13.0	.930	.957	.975	.986	.992	.996	.998	.999	1.000
14.0	.883	.923	.952	.971	.983	.991	.995	.997	.999
15.0	.819	.875	.917	.947	.967	.981	.989	.994	.997
16.0	.742	.812	.868	.911	.942	.963	.978	.987	.993
17.0	.655	.736	.805	.861	.905	.937	.959	.975	.985
18.0	.562	.651	.731	.799	.855	.889	.932	.955	.972
19.0	.469	.561	.647	.725	.793	.849	.893	.927	.951
20.0	.381	.470	.559	.644	.721	.787	.843	.888	.922
21.0	.302	.384	.471	.558	.640	.716	.782	.838	.883
22.0	.232	.306	.387	.472	.556	.637	.712	.777	.832
23.0	.175	.238	.310	.389	.472	.555	.635	.708	.772
24.0	.128	.180	.243	.314	.392	.473	.554	.632	.704
25.0	.092	.134	.185	.247	.318	.394	.473	.553	.629

Table I   Poisson Distribution: Selected Probabilities (continued)

np	c								
	27	*28*	*29*	*30*	*31*	*32*	*33*	*34*	*35*
7.0									
7.2									
7.4									
7.6									
7.8									
8.0									
8.5									
9.0									
9.5									
10.0									
10.5									
11.0									
11.5									
12.0									
13.0									
14.0	.999	1.000							
15.0	.998	.999							
16.0	.996	.998	.999	1.000					
17.0	.991	.995	.997	.999	.999	1.000			
18.0	.983	.990	.994	.997	.998	.999	1.000		
19.0	.969	.980	.988	.993	.996	.998	.999	.999	1.000
20.0	.948	.966	.978	.987	.992	.995	.997	.999	.999
21.0	.917	.944	.963	.976	.985	.991	.994	.997	.998
22.0	.877	.913	.940	.959	.973	.983	.989	.994	.996
23.0	.827	.873	.908	.936	.956	.971	.981	.988	.993
24.0	.768	.823	.868	.904	.932	.953	.969	.979	.987
25.0	.700	.763	.818	.863	.900	.929	.950	.966	.978

Table I   Poisson Distribution: Selected Probabilities (continued)

np	36	37	38	39	40	41	42
20.0	1.000						
21.0	.999	.999	1.000				
22.0	.998	.999	.999	1.000			
23.0	.996	.997	.999	.999	1.000		
24.0	.992	.995	.997	.998	.999	.999	1.000
25.0	.985	.991	.994	.997	.998	.999	.999

Index

A

Acceptable quality level (AQL), 220, 228, 231
Acceptance number, 219
Acceptance sampling, 12, 17, 215–253
    comparison of sampling plans, 239–243
    double sample sampling plans, 232–234
    inspection costs, 243–248
    OC curves for double sample sampling plans, 234–239
    operating characteristic curves, 220–221
    published sampling plans, 248–250
        ANSI/ASQC Z1.4-1993, 248–249
        ANSI/ASQC Z1.9-1993, 249–250
        Dodge-Romig tables, 250
    sampling probabilities, 221–231
    sampling risks, 219–220
Addition rule, 45
    for mutually exclusive events, 45
    for non-mutually exclusive events, 47
Adjusting factor, 97
Aging, of sample specimens, 33

Algebra
    format for stating hypothesis, 115
    language of, 45
Alternative hypothesis, 114, 123, 128, 130
American National Standards Institute (ANSI), 4
American Society for Quality (ASQ), 4
ANSI, see American National Standards Institute
AOQ, see Average outgoing quality
AOQL, see Average outgoing quality limit
AQL, see Acceptable quality level
Areas under normal curve, 74
α risk, 221
Arithmetic average, 33
Arithmetic mean, 32, 36, 39
ASN, see Average sample number
ASQ, see American Society for Quality
Assignable variation, 139, 140, 152
ATI, see Average total inspection
Attribute data, 184
Average, 32, 39
    calculation of, 247
    curve, 248, 249
    outgoing quality (AOQ), 239
        curves, 241
        limit (AOQL), 240
        values, 249
    process, 193, 194

315